VIEW ON VIRAL EMORRHAGIC FEVER (VHF)

ARIN BHATTACHARYA

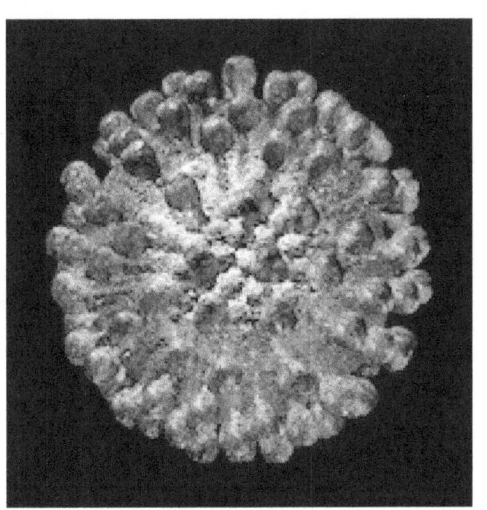

© Copyright is with the author

DEDICATION

This work is dedicated

to

My respected parents

I also thank god, my fellow mates

&

All the lovers of pharmacy world.

Acknowledgement

In any of scientific literally work many people work towards achievement of the desired goal i.e. publication of the research work in form book. I want to thanks my parents without their blessings, support and vision I could not have reached this stage.

Friendship is a treasured gift and fine friends are very few, my heartful thanks to my friend s that are always there for their support and positive views.

I would like to thanks the persons whom had criticize this project because it provided the positive impetus for completing the project in a more comprehensive and positive way.

Lastly it is only when someone writes a book that one realizes the true power of MSOffice software. It is simple- without this software this book would not be written.

Thank you Mr. Bill Gates and Microsoft Corp!

I strongly believe, "Success is blend of hard work and Destiny" and I think 'God' who have perpetually patronized me with consciousness and love to ladder the success. Thankful I ever remain....

Arin Bhattacharya

Table of Contents

CHAPTER NO.	NAME OF CHAPTER	PAGE NO
1	INTRODUCTION	6
2	BASICS OF RNA VIRUSES AND RNA VIRUSES FAMILIES RESPONSIBLE FOR VIRAL HEMORRHAGIC FEVER	7
3	LASSA FEVER	8
4	LUJO VIRUS	15
5	ARGENTINE HEMORRHAGIC FEVER(AHF)	17
6	BOLIVIAN HEMORRHAGIC FEVER (BHF)	20
7	BRAZILIAN HEMORRHAGIC FEVER (BZHF)	22
8	VENEZUELAN HEMORRHAGIC FEVER	24
9	HANTAVIRUS	26
10	CRIMEAN-CONGO HEMORRHAGIC FEVER(CCHF)	32
11	RIFT VALLEY FEVER	38
12	EBOLA VIRUS	45

13	MARBURG VIRUS	56
14	DENGUE	64
15	YELLOW FEVER	84
16	FLAVIVIRIDAE VIRUS TICK-BORNE ENCEPHALITIS	96
17	CONCLUSION	100
18	DIAGRAMS AND FIGURES	101

CHAPTER 1 INTRODUCTION

The viral hemorrhagic (or haemorrhagic) fevers (VHFs) are a diverse group of animal and human illnesses that may be caused by five distinct families of RNA viruses: the families Arenaviridae, Filoviridae, Bunyaviridae, Flaviviridae, and Rhabdoviridae. All types of VHF are characterized by fever and bleeding disorders and all can progress to high fever, shock and death in many cases. Some of the VHF agents cause relatively mild illnesses, such as the Scandinavian nephropathia epidemica, while others, such as the African Ebola virus, can cause severe, life-threatening disease. This work look in into microbiological parameters of the viruses causing viral hemorrhagic fever, diseases that come under the broad class of viral hemorrhagic fever , sign and symptoms of viral hemorrhagic fever, pathophysiology behind the viral hemorrhagic, management, and scope of research in this field

CHAPTER 2 BASICS OF RNA VIRUSES AND RNA VIRUSES FAMILIES RESPONSIBLE FOR VIRAL HEMORRHAGIC FEVERS

The viral hemorrhagic (or haemorrhagic) fevers (VHFs) are a diverse group of animal and human illnesses that may be caused by five distinct families of RNA viruses: the families Arenaviridae, Filoviridae, Bunyaviridae, Flaviviridae, and Rhabdoviridae.

To understand the hemorrhagic (or haemorrhagic) fevers (VHFs) we should first know the basics about the RNA viruses. An RNA virus is a virus that has RNA (ribonucleic acid) as its genetic material. This nucleic acid is usually single-stranded RNA (ssRNA), but may be double-stranded RNA (dsRNA).

Five families of RNA viruses have been recognised as being able to cause this syndrome.

- The family Arenaviridae include the viruses responsible for Lassa fever, Lujo virus, Argentine, Bolivian, Brazilian and Venezuelan hemorrhagic fevers.

- The family Bunyaviridae include the members of the Hantavirus genus that cause hemorrhagic fever with renal syndrome (HFRS), the Crimean-Congo hemorrhagic fever(CCHF) virus from the Nairovirus genus, Garissa virus from the Orthobunyavirus and the Rift Valley fever (RVF) virus from the Phlebovirus genus.

- The family Filoviridae includes Ebola virus and Marburg virus.

- The family Flaviviridae includes dengue, yellow fever, and two viruses in the tick-borne encephalitis group that cause VHF: Omsk hemorrhagic fever virus and Kyasanur Forest disease virus.

- In September 2012 scientists writing in the journal PLOS Pathogens reported the isolation of a member of the Rhabdoviridae responsible for 2 fatal and 2 non-fatal cases of hemorrhagic fever in the Bas-Congo district of the Democratic Republic of Congo. The non-fatal cases occurred in healthcare workers involved in the treatment of the other two, suggesting the possibility of person-to-person transmission. This virus appears to be unrelated to previously known Rhabdoviruses.

CHAPTER 3 LASSA FEVER

3.1 Introduction

Lassa fever or Lassa hemorrhagic fever (LHF) is an acute viral hemorrhagic fever caused by the Lassa virus.

It was first described in the town of Lassa, Nigeria in 1969. The infection is endemic in West African countries, and causes 300,000–500,000 cases annually, with approximately 5,000 deaths. Outbreaks of the disease have been observed in Nigeria, Liberia, Sierra Leone, Guinea, and the Central African Republic, but it is believed that human infections also exist in Democratic Republic of the Congo, Mali, and Senegal. Lassa fever is a viral hemorrhagic fever in West Africa. Studies show up to half a million cases of Lassa fever per year in West Africa, with 5000 resulting in death. Results Lassa virus was detected in 25 of 60 (42%) patients in northern and central Edo. The Lassa Virus effects adults and children alike; no matter your age you can be at risk for Lassa.

Presentation of cases used to be highest during the dry season (January to March) and lowest during the wet season (May to November). However, recent data from Kenema (1999-2002) show that admissions were highest during the change from the dry to the wet season . This change might be related partly to population movements during the civil unrest in Sierra Leone and overcrowding among refugees. Travel becomes increasingly difficult as the wet season progresses and may help to account for the decrease in numbers of cases later in the season. All the cases reported were diagnosed clinically. Until 1998 laboratory confirmation of diagnosis was available retrospectively and 60-70% of cases were confirmed, but in 2000 over half of a series of 22 cases were wrongly diagnosed. Thus, these recent apparent changes in infection patterns must be interpreted with caution.

People of all ages are susceptible. The disease is mild or has no observable symptoms in about 80% of people infected, but 20% have a severe multisystem disease. The incubation period is 6-21 days. The virus is excreted in urine for three to nine weeks from infection and in semen for three months. The extent of sexual transmission is unknown.

Sensorineural hearing deficit is a feature of the disease: it was found in 29% of confirmed cases compared with none of febrile controls in hospital inpatients. In the general population, 81% of those who experienced sudden deafness had antibodies to Lassa virus versus 19% of matched controls. There is no apparent relation between the severity of viral illness, initial hearing loss, or subsequent recovery. Lassa fever was responsible for 10-16% of all adult medical admissions in 1987 in two hospitals studied in Sierra Leone and for about 30% of adult deaths. The case fatality rate in Kenema varied from 12% to 23% for the period 1997-2002. A recent case series showed low admission rates and high case fatality rates for people aged less than 18 years (who make up 51% of the total population (United Nations Development Programme)) compared with older people. During pregnancy, high rates of maternal death (29%) and fetal and neonatal loss (87%) have been recorded (uterine evacuation improves outcome significantly), with 25% of all maternal deaths in Sierra Leone being due to Lassa fever. An estimate of the case fatality rate in the general population is 1-2%, much lower than in hospitalised cases, possibly as a consequence of differences in severity.

Using the figures for rural populations (available from the United Nations Development Programme) and the epidemiology of the disease, we estimate that the "at risk" seronegative population (in Sierra Leone, Guinea, and Nigeria) may be as high as 59 million, with an annual incidence of illness of three million, fatalities up to 67 000, and up to three million reinfections. Until a complete picture of Lassa fever is known, these are rough estimates. Comparable data are unavailable for the other countries where seropositivity has been recorded.

Lassa virus is zoonotic (transmitted from animals), in that it spreads to man from rodents, specifically multi-mammate rats (*Mastomys natalensis*).

This is probably the most common rodent in equatorial Africa, ubiquitous in human households and eaten as a delicacy in some areas In these rats infection is in a persistent asymptomatic state. The virus is shed in their excreta (urine and feces), which can be aerosolized. In fatal cases, Lassa fever is characterized by impaired or delayed cellular immunity leading to fulminant viremia.

Infection in humans typically occurs by exposure to animal excrement through the respiratory or gastrointestinal tracts. Inhalation of tiny particles of infective material (aerosol) is believed to be the most significant means of exposure. It is possible to acquire the infection through broken skin or mucous membranes that are directly exposed to infective material.

Transmission from person to person has also been established, presenting a disease risk for healthcare workers. Frequency of transmission via sexual contact has not been established.

After an incubation period of six to twenty-one days, an acute illness with multiorgan involvement develops. Non-specific symptoms include fever, facial swelling, and muscle fatigue, as well as conjunctivitis and mucosal bleeding. The other symptoms arising from the affected organs are:

Gastrointestinal tract

- Nausea
- Vomiting (bloody)
- Diarrhea (bloody)
- Stomach ache
- Constipation
- Dysphagia (difficulty swallowing)
- Hepatitis

Cardiovascular system

- Pericarditis
- Hypertension
- Hypotension
- Tachycardia (abnormally high heart rate)

Respiratory tract

- Cough
- Chest pain
- Dyspnoea
- Pharyngitis
- Pleuritis

Nervous system

- Encephalitis
- Meningitis
- Unilateral or bilateral hearing deficit
- Seizures

Clinically, Lassa fever infections are difficult to distinguish from other viral hemorrhagic fevers such as Ebola and Marburg, and from more common febrile illnesses such as malaria.

There is a range of laboratory investigations that are performed to diagnose the disease and assess its course and complications. ELISA test for antigen and IgM antibodies gives 88% sensitivity and 90% specificity for the presence of the infection. Other laboratory findings in Lassa fever include lymphopenia (low white blood cell count), thrombocytopenia (low platelets), and elevated aspartate aminotransferase (AST) levels in the blood. Lassa fever can also be found in cerebrospinal fluid.[15] In West Africa, where Lassa is most prevalent, it is difficult for doctors to diagnose due to the absence of proper equipment to perform tests..

Lassa fever could be done by early and aggressive treatment using Ribavirin.

3.2 Drug profile of Ribavirin

3.2.1 History of Ribavirin

Ribavirin (originally also known as Virazole) is a synthetic chemical not found in nature. It was first synthesized in 1970 at ICN Pharmaceuticals, Inc. (later Valeant Pharmaceuticals International) by chemist Joseph T. Witkowski, under the direction of laboratory director Roland K. Robins.

Ribavirin was discovered as part of a systematic ICN search of antiviral and antitumor activity in synthetic nucleosides. This was inspired in part by discovery (in the 1960s) of antiviral activity from naturally-occurring purine-like nucleoside antibiotics like showdomycin, coformycin, and pyrazomycin. These agents had too much toxicity to be clinically useful (and their antiviral activity may be incidental), but they served as the starting point for pharmaceutical chemists interested in antivirals and antimetabolic chemotherapeutic agents.

In 1972 it was reported that ribavirin was active against a variety of RNA and DNA viruses in culture and in animals, without undue toxicity. Ribavirin protected mice against mortality from both A and B strains of influenza, and ICN originally planned to market it as an anti-influenza drug. Results in human trials against experimental influenza infection were mixed, however, and the FDA ultimately did not approve this indication for ribavirin use in humans, thereby causing a severe financial shock to ICN.

Although ICN was allowed in 1980 to market ribavirin, in inhalant form, for RSV infection in children, the U.S. market for this indication was small. By the time oral ribavirin was finally approved by the FDA as part of a combination treatment (with interferon) for hepatitis C in 1998, the original ICN patents on ribavirin itself had expired, and (notwithstanding subsequent patent disputes) ribavirin had become essentially a generic drug.

Physically ribavirin is similar to the sugar D-ribose from which it is derived. It is freely soluble in water, and is re-crystallized as fine silvery needles from boiling methanol. The three free sugar hydroxyls make the pure drug hydrophilic enough that it is only sparingly soluble in anhydrous ethanol.

Classically, ribavirin is prepared from natural D-ribose by blocking the 2', 3' and 5' OH groups with benzyl groups, then derivatizing the 1' OH with an acetyl group which acts as a suitable leaving group upon nucleophilic attack. The ribose 1' carbon attack is accomplished with a 1,2,4 triazole-3-carboxymethyl ester, which directly attaches the 1' nitrogen of the triazole to the 1' carbon of the ribose, in the proper 1-β-D isomeric position. The bulky benzyl groups hinder attack at the other sugar carbons. Following purification of this intermediate, treatment with ammonia in methanolic conditions then simultaneously deblocks the ribose hydroxyls, and converts the triazole carboxymethyl ester to the carboxamide. Following this step, ribavirin may be recovered in good quantity by cooling and crystallization.

3.2.2 Pharmacology of Ribavirin

The proposed sites of action of the anti viral drugs are shown by the following figure

Mechanism of action the anti viral drugs

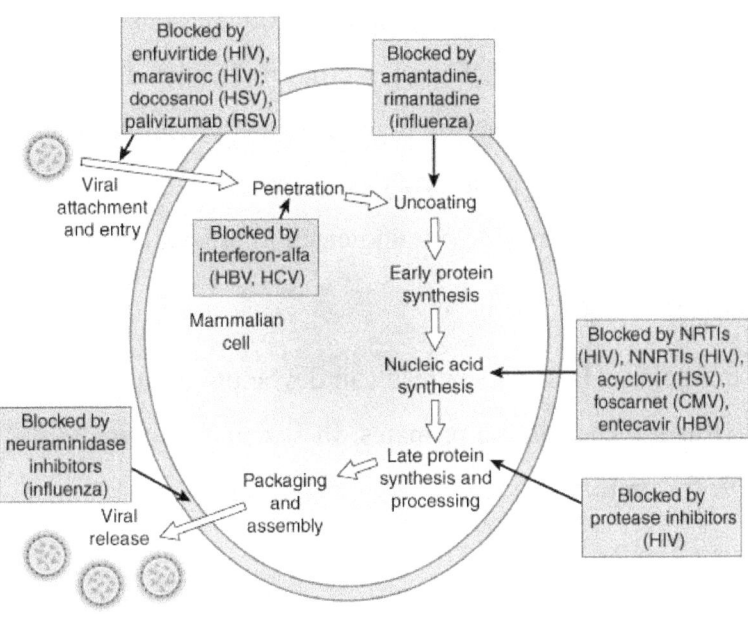

Ribavirin is a synthetic gunosine analog. It is effective against a broad spectrum of RNA and DNA viruses. For examples, ribavirin is used in treating infants and young children with severe RSV infection. Ribavirin is also effective in chronic hepatitis C infections when used in combination with interferons. Ribavirin may reduce the mortality and viremia of Lassa fever.

- Mode of action: The mode of action of ribavirin has been studied only for the influenza viruses. The drug is first converted to the 5'-phosphate derivatives, the major product being the compound ribavirin-triphosphate, which exerts its antiviral action by inhibiting guanosine triphosphate formation, preventing viral mRNA capping, and blocking RNA-dependent RNA polymerase. [Note: Rhinoviruses and enteroviruses, which contain preformed mRNA and do not need to synthesize mRNA in the host cell to initiate an infection, are relatively resistant to the action of ribavirin.]

- Pharmacokinetics: Ribavirin is effective orally and intravenously. Absorption is increased if the drug is taken with a fatty meal. An aerosol is used in certain respiratory viral conditions, such as the treatment of RSV infection. Studies of drug distribution in primates have shown retention in all tissues, except brain. The drug and its metabolites are eliminated in the urine

- Adverse effects: Side effects reported for oral or parenteral use of ribavirin have included dose-dependent transient anemia. Elevated bilirubin has been reported. The aerosol may be safer, although respiratory function in infants can deteriorate quickly after initiation of aerosol treatment. Therefore, monitoring is essential. Because of teratogenic effects in experimental animals, ribavirin is contraindicated in pregnancy

3.2.3 Vaccine for Lassa fever

Current research in the field of Lassa fever result in development of vaccine which is useful in treating Lassa fever virus infection in the primates. In Vaccine combination of Glycoprotein G1 and G2 are responsible for protective action.

CHAPTER 4 LUJO VIRUS

4.1 Introduction

Lujo is a bisegmented RNA virus — a member of the family Arenaviridae — and a known cause of viral hemorrhagic fever (VHF) in humans. Its name was suggested by the Special Pathogens Unit of the National Institute for Communicable Diseases of the National Health Laboratory Service (NICD-NHLS) by using the first two letters of the names of the cities involved in the 2008 outbreak of the disease, Lusaka (Zambia) and Johannesburg (Republic of South Africa). It is the second pathogenic arenavirus to be described from the African continent — the first being Lassa virus — and since 2012 has been classed as a "Select Agent" under U.S. law.

4.2 History of Lujo Virus

The discovery of this novel virus was described following a highly fatal nosocomial (hospital) outbreak of VHF in Johannesburg.

The first case was a female travel agent who lived in the outskirts of Lusaka. She suffered from fever which grew worse with time. She was evacuated to Johannesburg for medical treatment. Almost two weeks later, the paramedic that nursed the patient on the flight to South Africa, also fell ill and was also brought to Johannesburg for medical treatment. At this time the connection between these two patients was recognized by the attending physician in the Johannesburg hospital. Together with the NICD-NHLS the clinical syndrome of VHF was recognized and specimens from the second patient were submitted for laboratory confirmation.[2] In addition, a cleaner and a nurse that had contact with the first patient also fell ill. A second nurse was infected through contact with the paramedic. The outbreak had a high case fatality rate with 4 of 5 identified cases resulting in death.

The Special Pathogens Unit of the NICD-NHLS together with colleagues from the Special Pathogens Unit of the U.S. Centers for Disease Control and Prevention (CDC) identified the etiological agent of the outbreak as an Old World arenavirus using molecular and serological tests. Sequencing and phylogenetic investigation of partial genome sequencing indicated that this virus was not Lassavirus and likely a previously unreported arenavirus. This was corroborated by full genome sequencing that was conducted by the NICD-NHLS, CDC and collaborators from Columbia University in New York.

4.3 Distribution of Lujo Virus

The distribution of this newly described arenavirus is uncertain. To date this virus has only been reported from a patient from Zambia and a subsequent nosocomial outbreak in South Africa.

4.4 Phylogenetics

Sequencing of the viral genome has shown that this virus belongs to the Old World arenavirus group. Comparisons with other viral genome sequences showed that this virus is equidistant from other Old World and New World arenaviruses. It is distantly similar to the other pathogenic African arenavirus, Lassa fever virus.

4.5 Clinical cases

This virus has been associated with an outbreak of five cases of VHF in September and October 2008. In four cases (80% of total known infections) the infection was fatal. The fifth case was treated with ribavirin early after onset of clinical disease (was detected through active contact tracing), an antiviral drug which is effective in treating Lassa fever, and survived; however, ribavirin's effectiveness against Lujo virus remains unknown.

CHAPTER 5 ARGENTINE HEMORRHAGIC FEVER(AHF)

Argentine hemorrhagic fever (AHF) or O'Higgins disease, also known in Argentina as *mal de los rastrojos*, stubble disease, is a hemorrhagic fever and zoonotic infectious disease occurring in Argentina. It is caused by the *Junín virus* (an arenavirus, closely related to the *Machupo virus*, causative agent of Bolivian hemorrhagic fever). Its vector is a species of rodent, the corn mouse.

5.1 Junin Virus

The Junin virus virion is enveloped with a variable diameter of between 50 and 300 nm. The surface of the particle encompasses a layer of T-shaped glycoprotein extensions, extending up to 10 nm from the envelope, which are important for mediating attachment and entry into host cells.

The Junin virus genome comprises two single stranded RNA molecules, each encoding two different genes in an ambisense orientation. The two segments are termed 'short (S)' and 'long (L)' due to their respective lengths. The short segment (around 3400 nucleotides in length) encodes the nucleocapsid protein and the glycoprotein precursor (GPC). The GPC is subsequently cleaved to form two viral glycoproteins, GP1 and GP2 which ultimately form the T-shaped glycoprotein spike which extends from the viral envelope. . The long segment (around 7200 nucleotides in length) encodes the viral polymerase and a zinc binding protein. It is spread by rodents

The list of signs and symptoms mentioned in various sources for Junin virus includes the 16 symptoms listed below:
- Enlarged lymph glands
- Hemorrhage
- Edema
- Congestion
- Leukopenia
- Thrombocytopenia

- Fever
- Fatigue
- Dizziness
- Aching muscles
- Weakness
- Exhaustion
- Bleeding under the skin
- Bleeding from internal organs
- Seizures
- Delirium

5.2 Clinical aspects

AHF is a grave acute disease which may progress to recovery or death in 1 to 2 weeks. The incubation time of the disease is between 10 and 12 days, after which the first symptoms appear: fever, headaches, weakness, loss of appetite and will. These intensify less than a week later, forcing the infected to lie down, and producing stronger symptoms such as vascular, renal, hematological and neurological alterations. This stage lasts about 3 weeks, untreated; the mortality of AHF reaches 15–30%. The specific treatment includes plasma of recovered patients, which, if started early, is extremely effective and reduces mortality to 1%.

5.3. Treatment

Ribavirin has also shown some promise in treating arenaviral diseases.

Candid no 1 is the vaccine used in the treatment of Argentine hemorrhagic fever (AHF).

McKee, Jr et al had evaluated the protective efficacy of Candid No. I. a live-attenuated vaccine against Argentine hemorrhagic fever(AHF) in non human primates. In the study Twenty rhesus macaques immunized 3 months previously with graded doses of Candid No. 1(16- 127. 000 PFU), as well as 4 placebo-inoculated controls, were challenged with 4.41 log,0 PFU of virulent P3790 strain Junin virus.

All controls developed severe clinical disease: 3 of 4 died. In contrast, all vaccinated animals were fully protected-. none developed any signs of AHF during a 105-day follow-up period.

Viremia and virus shedding were readily detected in all placebo vaccinated controls, while virus could be recovered only once (by amplification) from throat swabs of 2 Candid No. I vacci Key Words nees on day 21. Vigorous secondary-type neutralizing and im Junin virus munofluorescent antibody responses were seen in most vacciCandid No. I nees that had received 3 $\log_{10}$ PFU Candid No. I or fewer: all Argentine hemorrhagic fever others, including those receiving 127,200 PFU, maintained rela Viral hemorrhagic fever relatively stable titers during follow-up. Candid No. I was highly immunogenic and fully protective against lethal Junin virus

CHAPTER 6 BOLIVIAN HEMORRHAGIC FEVER (BHF)

6.1 Introduction

Bolivian hemorrhagic fever (BHF), also known as black typhus or Ordog Fever, is a hemorrhagic fever and zoonotic infectious disease originating in Bolivia after infection by Machupo virus.[1]

BHF was first identified in 1959 by a research group led by Karl Johnson an ambisense RNA virus of the Arenaviridae family. The mortality rate is estimated at 5 to 30 percent. Due to its pathogenicity, Machupo virus requires Biosafety Level Four conditions, the highest level.

In February and March 2007, some 20 suspected BHF cases (3 fatal) were reported to the El Servicio Departamental de Salud (SEDES) in Beni Department, Bolivia, and in February 2008, at least 200 suspected new cases (12 fatal) were reported to SEDES. In November 2011, a SEDES expert involved in a serosurvey to determine the extent of Machupo virus infections in the Department after the discovery of a second confirmed case near the departmental capital of Trinidad in November, 2011, expressed concern about expansion of the virus' distribution outside the endemic zone in Mamoré and Iténez provinces

6.2 Vector of Bolivian hemorrhagic fever (BHF)

The vector is the vesper mouse *Calomys callosus*, a rodent indigenous to northern Bolivia. Infected animals are asymptomatic and shed the virus in excreta, thereby infecting humans. Evidence of person-to-person transmission of BHF exists but is believed to be rare

6.3 Symptoms

The list of signs and symptoms mentioned in various sources for Bolivian hemorrhagic fever includes the symptoms listed below:
- Fever
- Headache
- Weakness
- Malaise
- Vomiting

- Muscle pain
- Loss of appetite
- Kidney symptoms
- Reduced blood pressure
- Bleeding gums
- Bleeding nose
- Petechiae
- Dehydration
- Flushed face
- Reduced urination frequency
- Reduced heart rate
- Neurological symptoms
- Vomiting blood
- Black tarry stool
- Gastrointestinal hemorrhage
- Blood in urine
- Delirium
- Alopecia

6.4 Treatment

Intravenous Ribavirin may be one of the treatments for Bolivian hemorrhagic fever (BHF). Kilgore et al. had given evidence regarding the efficiency of Intravenous Ribavirin in the research paper named "Treatment of Bolivian Hemorrhagic Fever with Intravenous Ribavirin" published in CID in 1997. The author finds that the clinical and laboratory data obtained during the treatment of patient shows that the intravenous Ribavirin may be used in Bolivian hemorrhagic fever (BHF), but there is scope of research to find which antivirals could be used for Bolivian hemorrhagic fever (BHF),

CHAPTER 7 BRAZILIAN HEMORRHAGIC FEVER (BZHF)

7.1 Introduction

Brazilian hemorrhagic fever (BzHF) is an infectious disease caused by the **Sabiá virus**, an Arenavirus. The Sabiá virus is an enveloped RNA virus and is highly infectious and lethal.

The incubation period is between 7–16 days, during which signs and symptoms can develop. Initial signs and symptoms can include fever, eye redness, fatigue, dizziness, muscle aches, loss of strength, and exhaustion. Severe cases show signs of bleeding under the skin, internal organs, or from body orifices like the mouth, eyes, or ears. Severely ill patients show shock, nervous system malfunction, coma, delirium, and seizures

The Sabiá virus can be acquired through inhalation, ingestion, the eyes, and contact with urine, saliva, blood, or feces of rodents.

To date, there have only been three reported infections of the Sabiá virus. Only one known case of naturally contracted Sabiá virus occurred in a woman staying in the village of Sabiá, outside of São Paulo, in 1990. Two other cases occurred in a clinical setting which are the viriologist who was responsible for the study of the woman's disease, contracted the disease as during the course of his research; he, fortunately, survived. Four years later, while working under level 3 biohazard conditions, a researcher at the Tropical Medicine Clinic at Yale-New Haven Hospital was exposed to the virus. Exposure apparently resulted when a centrifuge bottle containing infected tissue cracked and leaked into the spinning centrifuge, releasing aerosolized virus particles into the air. One of the scientists who were infected was treated with ribavirin.

7.2 Transmission

Like other New World arenaviruses, transmission is assumed to be via aerosolized virus particles. Close contact with infected individuals or suspected animal reservoirs or vectors are key factors in Sabia diagnosis. Though the animal reservoir is as yet unknown, a rodent found throughout the region surrounding the small village of Sabia has been implicated. In the history of the virus, laboratory-related infection has been a primary method of transmission, and therefore necessitates the utmost caution when handling the virus in a laboratory setting.

7.3 Symptoms

Due to the small number of individuals infected with Sabia virus, future cases may elaborate upon current known symptoms. Fever, headache, myalgia, nausea, vomiting, weakness, and pronounced sore throat were symptoms exhibited in all cases of Sabia infection. Additional symptoms include conjunctivitis, diarrhea, epigastric pain, and bleeding gums. In both cases that occurred in 1990, symptoms lasted approximately 15 days. However, symptoms ceased in the index case due to death, while the laboratory technician convalesced following this time period. Leucopenia, thrombocytopenia, and proteinuria were all present in each case, though these are hardly unique to the virus. Gastro-intestinal hemorrhage was marked in the index case, though generalized hemorrhagic fever appears to have been exhibited in all documented cases.

7.4 Treatment

Like other arenaviruses, Sabiá virus proved to be responsive to treatment with ribavirin. In confirmed cases of Sabiá infection, ribavirin is unquestionably the most appropriate treatment. Additionally, treatment of symptoms related to dehydration and bleeding is also recommended. Hemorrhage is most often the primary concern, meaning that fluid intake should be monitored carefully to compensate for vascular leaking and edema.

CHAPTER 8 VENEZUELAN HEMORRHAGIC FEVER

8.1 Introduction

Venezuelan hemorrhagic fever (VHF) is a zoonotic human illness first identified in 1989. The disease is most prevalent in several rural areas of central Venezuela and is caused by the Guanarito (GTOV) arenavirus belonging to the Arenaviridae family. The short-tailed cane mouse (Zygodontomys brevicauda) is the main host for GTOV which is spread mostly by inhalation of aerosolized droplets of saliva, respiratory secretions, urine, or blood from infected rodents. Person-to-person spread is possible, but uncommon.

8.2 Zygodontomys brevicauda

Zygodontomys brevicauda, also known as the Short-tailed Zygodont, Short-tailed Cane Mouse, or Common Cane Mouse, is a species of rodent in the genusZygodontomys of tribe Oryzomyini.

It occurs from Costa Rica via Panama, Colombia andVenezuela into Guyana, Suriname, French Guiana and northern Brazil, including Trinidad and Tobago. It includes three subspecies: Zygodontomys brevicauda brevicauda, Zygodontomys brevicauda cherriei, and Zygodontomys brevicauda microtinus. Many Zygodontomys brevicauda serve as viral reservoirs, causing illnesses such as Venezuelan hemorrhagic fever. _

8.3 History

From September 1989 through December 2006, the State of Portuguesa recorded 618 cases of VHF. Nearly all of the cases were individuals who worked or lived in Guanarito during the time they became infected. The case fatality rate was 23.1%.

Because the virus is contracted by aerosol dissemination, concern arose shortly after the first cases emerged in 1989 due to fear of biological warfare. Potential biological terrorism agents were identified and categorized in 1999 by the Centers for Disease Control and Prevention (CDC) as part of the Congressional initiative to further response capabilities to biological weapons . Arenaviruses causing hemorrhagic fevers, along with a genus of virus called filoviruses, were categorized in Category A; these are pathogens with the highest potential impact on public health safety.

A notable event in the timeline of this virus' scientific knowledge was the unexplained disappearance of a vial of the virus at the University of Texas Medical Branch Galveston National Laboratory, announced 2013 March 24.

8.4 Virus

Arenaviruses are enveloped, single-stranded, bisegmented RNA viruses with ambiences genomes. Based on their antigenic properties, arenaviruses have been classified into two major groups: the Old World arena viruses, and the New World arenaviruses. Old World arena viruses include lymphocytic choriomeningitis virus, and Lassa virus. New world arena viruses are further broken down into three clades, A, B, and C. The Guanarito arena virus belongs to clade B and is the cause of VHF. On the biosafety level scale of one to four, with four causing the most risk, the viruses causing hemorrhagic fevers have been assigned a four by the CDC.

8.5 Treatment

VHF has many similarities to Lassa fever and to the arenavirus hemorrhagic fevers that occur in Argentina and Bolivia. It causes fever and malaise followed by hemorrhagic manifestations and convulsions. Some presentations of the virus are also characterized by vascular damage, bleeding diathesis, fever, and multiple organ involvement. Clinical diagnosis of VHF has proven to be difficult based on the nonspecific symptoms. The disease is fatal in 30% of cases and is endemic to Portuguesa state and Barinas state in Venezuela.

Treatment and prevention for the VHF virus are limited and there are currently no licensed vaccines available that can act to prevent the disease. However, once infected, Ribavirin, an anti-viral drug given intravenously, is one way to treat VHF.

CHAPTER 9 HANTAVIRUS

9.1 Introduction

Hantaviruses are negative sense RNA viruses in the Bunyaviridae family. Humans may be infected with Hantaviruses through urine, saliva or contact with rodent waste products. Some Hantaviruses cause potentially fatal diseases in humans, such as hemorrhagic fever with renal syndrome (HFRS) and Hantavirus pulmonary syndrome (HPS), but others have not been associated with known human disease.

Human infections of Hantaviruses have almost entirely been linked to human contact with rodent excrement, but recent human-to-human transmission has been reported with the Andes virus in South America.[1] The name *hantavirus* is derived from the Hantan Riverarea in South Korea, which provided the founding member of the group: Hantaan virus (HTNV), isolated in the late 1970s by Ho-Wang Lee, Karl M. Johnson and his colleagues. Hantaan, Dobrava, Saaremaa, Seoul, and Puumala are several hantaviruses species that cause HFRS, also known as Korean hemorrhagic fever, epidemic hemorrhagic fever and nephropathis epidemica.

9.2 History

The hantaviruses are a relatively newly discovered genus of viruses. An outbreak of Korean Hemorrhagic Fever among American and Korean soldiers during the Korean War (1951-1953) was later found to be caused by a hantavirus infection. More than 3000 troops became ill with symptoms that included renal failure, generalized hemorrhage, and shock. It had a 10% mortality rate. This outbreak sparked a 25-year search for the etiologic agent. Ho-Wang Lee, a South Korean virologist, Karl M. Johnson, an American tropical virologist, and his colleagues isolated Hantaan virus in 1976 from the lungs of striped field mice. In late Medieval England a mysterious sweating sickness swept through the country in 1485 just before the Battle of Bosworth Field. Noting the similar symptoms which overlap with Hantavirus pulmonary syndrome , several scientists have theorised that the virus may have been the cause of the disease. The hypothesis was criticised because sweating sickness was recorded as being transmitted human-to-human whereas hantaviruses were not known to spread in this way. Limited transmission via human-to-human contact has since been shown in Hantavirus outbreaks in Argentina.

In 1993, an outbreak of Hantavirus pulmonary syndrome occurred in the Four Corners region in the southwestern United States. The viral cause of the disease was found only weeks later and was called the Sin Nombre virus (SNV, in Spanish, "Virus sin nombre", for "nameless virus"). Its rodent host, the Deer mouse (*Peromyscus maniculatus*), was first identified by Terry Yates, a professor at the University of New Mexico.

9.3 Clinical syndromes

9.3.1 Hemorrhagic fever with renal syndrome

Hemorrhagic fever with renal syndrome (HFRS) is a group of clinically similar illnesses caused by species of hantaviruses from the family Bunyaviridae. It is also known as Korean hemorrhagic fever, epidemic hemorrhagic fever, and nephropathis epidemica. The species that cause HFRS include Hantaan, Dobrava, Saaremaa, Seoul, and Puumala.[6] It is found in Europe, Asia, and Africa.

In hantavirus induced hemorrhagic fever, incubation time is between two to four weeks in humans before symptoms of infection present. Severity of symptoms depends on the viral load.

The course of the illness can be split into five phases:

- **Febrile phase**: Symptoms include redness of cheeks and nose, fever, chills, sweaty palms, diarrhea, malaise, headaches, nausea, abdominal and back pain, respiratory problems such as the ones common in the influenza virus, as well as gastro-intestinal problems. These symptoms normally occur for three to seven days and arise about two to three weeks after exposure.

- **Hypotensive phase**: This occurs when the blood platelet levels drop and symptoms can lead to tachycardia and hypoxemia. This phase can last for 2 days.

- **Oliguric phase**: This phase lasts for three to seven days and is characterised by the onset of renal failure and proteinuria.

- **Diuretic phase**: This is characterized by diuresis of three to six litres per day, which can last for a couple of days up to weeks.

- **Convalescent phase**: This is normally when recovery occurs and symptoms begin to improve.

This syndrome can also be fatal. In some cases, it has been known to cause permanent renal failure.

9.3.2 Hantavirus pulmonary syndrome

Hantavirus pulmonary syndrome (HPS) is found in the United States and Central and South America. It is an often fatal pulmonary disease. In the United States, the causative agent is Sin Nombre virus carried by deer mice. Prodromal symptoms include flu-like symptoms such as fever, cough, myalgia, headache, and lethargy. It is characterized by a sudden onset of shortness of breath with rapidly evolving pulmonary edema that is often fatal despite mechanical ventilation and intervention with potent diuretics. It has a fatality rate of 60%.

Hantavirus pulmonary syndrome was first recognized during the 1993 outbreak in the Four Corners region of the southwestern United States. It was identified by Dr. Bruce Tempest. It was originally called "Four Corners disease," but the name was changed to Sin Nombre virus after complaints by Native Americans that the name 'Four Corners' stigmatized the region It has since been identified throughout the United States. Rodent control in and around the home remains the primary prevention strategy.

9.4 Epidemiology

Regions especially affected by hemorrhagic fever with renal syndrome include China, the Korean Peninsula, Russia (Hantaan, Puumala and Seoul viruses), and northern and western Europe (Puumala and Dobrava virus). Regions with the highest incidences of hantavirus pulmonary syndrome include Patagonia Argentina, Chile, Brazil, the United States, Canada, and Panama.

Asia

In China, the Korean Peninsula, and Russia, hemorrhagic fever with renal syndrome is caused by Hantaan, Puumala and Seoul viruses. Korean Hemorrhagic Fever was first documented during the Korean War when American troops became ill.

South America

The two agents of HPS in South America are Andes virus (also called Oran, Castelo de Sonhos, Lechiguanas, Juquitiba, Araraquara, and Bermejo viruses, among many other synonyms), which is the only hantavirus that has shown (albeit uncommonly) an interpersonal form of transmission, and Laguna Negra virus, an extremely close relative of the previously-known Rio Mamore virus.

Rodents that have been shown to carry Hantaviruses include Abrothrix longipilis and Oligoryzomys longicaudatus.

North America

In the U.S., minor cases of HPS include Sin Nombre virus, New York virus, Bayou virus, and possibly Black Creek Canal virus.

In the United States, as of July 2010 eight states had reported 30 or more cases of Hantavirus since 1993

In late August and early September of 2012, eight new cases of Hantavirus were confirmed, including three deaths, in the Curry Village area of Yosemite National Park.

In addition to infecting rodents hantaviruses are known to be carried by shrews (order *Soricomorpha*, family *Soricidae*) and moles (family *Talpidae*).[1]

In Mexico a number of rodents have been found to carry hantaviruses: *Megadontomys thomasi*, *Neotoma picta*, *Peromyscus beatae*, *Reithrodontomys megalotis* and *Reithrodontomys sumichrasti*.

In Canada, one confirmed death in Northern B.C in Jan 2013.

Europe

In Europe three hantaviruses - Puumala, Dobrava and Saaremaa viruses - are known to cause haemorrhagic fever with renal syndrome. Puumala usually causes a generally mild disease - nephropathia epidemica - which typically presents with fever, headache, gastrointestinal symptoms, impaired renal function and blurred vision. Dobrava infections while similar often also have haemorrhagic complications.

There are few reports of confirmed Saaremaa infections but these appear to be similar to those caused by Puumala and less pathogenic than Dobrava.

Puumala is carried by its rodent host, the bank vole (*Clethrionomys glareolus*) and is present through most of Europe excluding the Mediterranean region. Dobrava and Saaremaa carried respectively by the yellow necked mouse (*Apodemus flavicollis*) and the striped field mouse (*Apodemus agrarius*) are reported mainly in eastern and central Europe.

Africa

A hantavirus (Sangassou virus) has been isolated in Africa that causes hemorrhagic fever with renal syndrome

9.5 Virology

Classification

Hantaviruses are Bunyaviruses. The Bunyaviridae family is divided into five genera.

1) Orthobunyavirus,

2) Nairovirus,

3) Phlebovirus,

4) Tospovirus,

5) Hantavirus.

Like all members of this family, hantaviruses have genomes comprising three negative-sense, single-stranded RNA segments, and so are classified as negative sense RNA viruses. [Negative-sense (3' to 5') viral RNA is complementary to the viral mRNA and thus must be converted to positive-sense RNA by an RNA polymerase prior to translation. Negative-sense RNA (like DNA) has a nucleotide sequence complementary to the mRNA that it encodes. Like DNA, this RNA cannot be translated into protein directly. Instead, it must first be transcribed into a positive-sense RNA that acts as an mRNA.

9.6 Life cycle

Entry into host cells is thought to occur by attachment of virions to cellular receptors and subsequent endocytosis. Nucleocapsids are introduced into the cytoplasm by pH-dependent fusion of the virion with the endosomal membrane. Subsequent to release of the nucleocapsids into cytoplasm, the complexes are targeted to the ER-Golgi Intermediate compartments (ERGIC) through microtubular associated movement resulting in the formation of viral factories at ERGIC.

These factories then facilitate transcription and subsequent translation of the viral proteins. Transcription of viral genes must be initiated by association of the L protein with the three nucleocapsid species. In addition to transcriptase and replicase functions, the viral L protein is also thought to have an endonuclease activity that cleaves cellular messenger RNAs (mRNAs) for the production of capped primers used to initiate transcription of viral mRNAs. As a result of this "cap snatching," the mRNAs of hantaviruses are capped and contain nontemplated 5' terminal extensions.

The G1 (aka Gn) and G2 (Gc) glycoproteins form hetero-oligomers and are then transported from the endoplasmic reticulum to the Golgi complex, where glycosylation is completed. The L protein produces nascent genomes by replication via a positive-sense RNA intermediate. Hantavirus virions are believed to assemble by association of nucleocapsids with glycoproteins embedded in the membranes of the Golgi, followed by budding into the Golgi cisternae. Nascent virions are then transported in secretory vesicles to the plasma membrane and released by exocytosis.

9.7 Treatment and prevention

There is no known antiviral treatment, but natural recovery from the virus is possible with supportive treatment. Patients with suspected hantavirus are usually admitted to the hospital and given oxygen and mechanical ventilation support to help them breathe during the acute pulmonary stage. As the virus can be transmitted by rodent saliva, excretia, and bites, control of rats and mice in areas frequented by humans is key for disease prevention.

General prevention can be accomplished by disposing of rodent nests, sealing any cracks and holes in homes where mice or rats could get in, setting up traps, laying down poisons or using natural predators such as cats in the home.

CHAPTER 10   CRIMEAN-CONGO HEMORRHAGIC FEVER(CCHF)

10.1 Introduction

Crimean–Congo hemorrhagic fever (CCHF) is a widespread tick-borne viral disease, a zoonosis of domestic animals and wild animals, that may affect humans. The pathogenic virus, especially common in East and West Africa, is a member of the Bunyaviridae family of RNA viruses. Clinical disease is rare in infected mammals, but commonly severe in infected humans, with a 30% mortality rate. Outbreaks of illness are usually attributable to handling infected animals or people.

Crimean-Congo hemorrhagic fever is one of the most widely distributed viral hemorrhagic fevers. This disease occurs in much of Africa, the Middle East and Asia, as well as parts of Europe. Changes in climatic conditions could expand the range of its tick vectors, and increase the incidence of disease. The CCHF virus is also a potential bioterrorist agent; it has been listed in the U.S. as a CDC/NIAID Category C priority pathogen

10.2 History

A case reported in the 12th century of a hemorrhagic disease from what is now Tajikistan may have been the first known case of Crimean–Congo hemorrhagic fever. Soviet scientists first identified the disease they called Crimean hemorrhagic fever in 1944 and established its viral etiology by passage of the virus through human "volunteers" (fatality rate unreported), but were unable to isolate the agent at that time. In June 1967, Soviet virologist Mikhail Chumakov registered an isolate from a fatal case that occurred in Samarkand (on the ancient Silk Road in Central Asia, not the Crimea) in the Catalogue of Arthropod-borne Viruses. Four months earlier, virologists Jack Woodall, D Simpson and others had published initial reports on a virus they called the Congo virus, first isolated in 1956 by physician Ghislaine Courtois, head of the Provincial Medical Laboratory, Stanleyville, Belgian Congo. Strain V3010, isolated by Courtois, was sent to the Rockefeller Foundation Virus Laboratory (RFVL) in New York City and found to be identical to another strain from Uganda, but to no other named virus at that time. Chumakov later sent his strain to the RFVL, where it was found to be identical to the Congo virus.

The International Committee on Taxonomy of Viruses proposed the name Congo-Crimean hemorrhagic fever virus, but the Soviets insisted on Crimean–Congo hemorrhagic fever virus. Against all principles of scientific nomenclature based on priority of publication, it was adopted as the official name in 1973 in possibly the first instance of a virus losing its name to politics and the Cold War. However, since then Congo-Crimean or just Congo virus has been used in many reports, which would be missed in searches of medical databases using the official name. These reports include records of the occurrence of the virus or antibodies to the virus from Greece, Portugal, South Africa, Madagascar (the first isolation from there), the Maghreb, Dubai, Saudi Arabia, Kuwait and Iraq.

10.3 The Etiological Agent

Crimean–Congo hemorrhagic fever (CCHF) as mentioned before is caused by a virus. This virus is a member of the Nairovirus genus of the family Bunyaviridae, Hantavirus also belongs to the same family. Nairovirus genus contains 7 species of virus with 34 strains reported till date.

All of these viruses are transmitted by either ixodid or argasid ticks (i.e., hard or soft ticks, respectively). Structurally the Crimean–Congo hemorrhagic fever (CCHF) virus is an rNA virus, it is spherical, approximately 100 nm in diameter, and has a host cell-derived lipid bilayered envelope.

10.4 Geographic Distribution

CCHFV is widespread in Africa, the Middle East and Asia. It has also been found in parts of Europe including southern portions of the former USSR (Crimea, Astrakhan, Rostov, Uzbekistan, Kazakhstan, Tajikistan), Turkey, Bulgaria, Greece, Albania and Kosovo province of the former Yugoslavia. Limited serological evidence suggests that CCHFV might also occur in parts of Hungary, France and Portugal. The occurrence of this virus is correlated with the distribution of *Hyalomma* spp., the principal tick vectors. According to a WHO report in 2008 CCHFV evidences were found in country like India also.

10.5 Transmission

CCHFV usually circulates between asymptomatic animals and ticks in an enzootic cycle. This virus has been found in at least 31 species of ticks, including seven genera of the family Ixodidae (hard ticks). Members of the genus *Hyalomma* seem to be the principal vectors. Transovarial, transstadial and venereal transmission occur in this genus. *Hyalomma marginatum marginatum* is particularly important as a vector in Europe, but CCHFV is also found in *Hyalomma anatolicum anatolicum* and other *Hyalomma* spp.

Other ixodid ticks including members of the genera Rhipicephalus, *Boophilus*, *Dermacentor* and *Ixodes* may also transmit the virus locally. Although CCHFV has been reported in other families of invertebrates, these species may not be biological vectors; the virus may have been ingested in a recent blood meal.

In one study, CCHFV was reported from a biting midge (*Culicoides* spp.). It has also been found in two species of Argasidae (soft ticks); however, experimental infections suggest that CCHFV does not replicate in this family of ticks.

Many species of mammals can transmit CCHFV to ticks when they are viremic. Small vertebrates such as hares and hedgehogs, which are infested by immature ticks, may be particularly important as amplifying hosts. With a few exceptions, birds seem to be refractory to infection; however, they may act as mechanical vectors by transporting infected ticks. Migratory birds might spread the virus between distant geographic areas.

Humans become infected through the skin and by ingestion. Aerosol transmission was suspected in a few cases in Russia. Sources of exposure include being bitten by a tick, crushing an infected tick with bare skin, contacting animal blood or tissues and drinking unpasteurized milk. Human-to-human transmission occurs, particularly when skin or mucous membranes are exposed to blood during hemorrhages or tissues during surgery. CCHFV is stable for up to 10 days in blood kept at 40°C (104°F). Possible horizontal transmission has been reported from a mother to her child.

10.6 Diagnostic Tests

Crimean-Congo hemorrhagic fever can be diagnosed by isolating CCHFV from blood, plasma or tissues. At autopsy, the virus is most likely to be found in the lung, liver, spleen, bone marrow, kidney and brain. CCHFV can be isolated in a variety of cell lines including SW-13, Vero, LLC-MK2 and BHK-21 cells. Cell cultures can only detect high concentrations of the virus, and this technique is most useful during the first five days of illness. Animal inoculation into newborn mice is more sensitive than culture, and can detect the virus for a longer period. CCHFV is identified by indirect immunofluorescence or reverse transcription-polymerase chain reaction (RT-PCR) assays. Virus isolation must be carried out in maximum biocontainment laboratories (BSL-4).

Crimean-Congo hemorrhagic fever is often diagnosed by RT-PCR on blood samples. This technique is highly sensitive. However, due to the genetic variability in CCHFV strains, a single set of primers cannot detect all virus variants, and most RT-PCR assays are either designed to detect local variants or lack sensitivity.

A real-time RT-PCR assay that can detect numerous variants has recently been published. Viral antigens can be identified with enzyme-linked immunoassay (ELISA) or immunofluorescence, but this test is less sensitive than PCR.

Crimean-Congo hemorrhagic fever can also be diagnosed by serology. Tests detect CCHFV-specific IgM, or a rise in IgG titers in paired acute and convalescent sera. IgG and IgM can usually be found with indirect immunofluorescence or ELISA after 7-9 days of illness. Other serologic tests such as complement fixation and hemagglutination inhibition were used to diagnose Crimean-Congo hemorrhagic fever in the past, but lacked sensitivity. In fatal cases, patients generally die without developing antibodies.

10.7 Treatment and Prevention

Treatment is mainly supportive. Ribavirin is used in some cases. Observational studies in humans and studies in experimentally infected mice support the use of this drug; however, no randomized human clinical trials have been published. Passive immunotherapy with hyperimmune serum has been tested in a few cases, but the value of this treatment is controversial.

In endemic regions, prevention depends on avoiding bites from infected ticks and contact with infected blood or tissues. Measures to avoid tick bites include tick repellents, environmental modification (brush removal, insecticides), avoidance of tick habitat and regular examination of clothing and skin for ticks. Clothing should be chosen to prevent tick attachment; long pants tucked into boots and long-sleeved shirts are recommended. Acaricides can be used on livestock and other domesticated animals to control ticks, particularly before slaughter or export.

Protective clothing and gloves should be worn whenever skin or mucous membranes could be exposed to viremic animals, particularly when blood and tissues are handled.

Unpasteurized milk should not be drunk. In meat, CCHFV is usually inactivated by post-slaughter acidification. It is also killed by cooking.

Strict universal precautions are necessary when caring for human patients. These recommendations include barrier nursing, isolation and the use of gloves, gowns, face-shields and goggles with side shields.

Prophylactic treatment with ribavirin has occasionally been used after high-risk exposures. Safe burial practices, including the use of 1:10 liquid bleach solution as a disinfectant, have been published. Laboratory workers must follow stringent biosafety precautions.

An inactivated vaccine from mouse brains has been used in the former Soviet Union and Bulgaria. In most countries, no vaccine is available.

10.8 Morbidity and Mortality

Climatic factors can influence the numbers of ticks in the environment and the incidence of disease. In some countries, Crimean-Congo hemorrhagic fever tends to be seasonal. This disease is most common in Iran during August and September, and in Pakistan from March to May and August to October.

Most cases are the result of occupational exposure. CCHF is particularly common in farmers, shepherds, veterinarians, abattoir workers and laboratory workers. Healthcare workers are also at high risk, particularly after exposure to patients' blood. During one nosocomial outbreak at a hospital in South Africa, 33% of medical personnel exposed via needlestick injuries became ill. Approximately 9% of those who had other forms of contact with infected blood also developed CCHF. In the general public, activities that increase tick exposure such as hiking and camping increase the risk of infection.

The average case fatality rate is 30-50%, but mortality rates from 10% to 80% have been reported in various outbreaks. The mortality rate is usually higher for nosocomial infections than after tick bites; this may be related to the virus dose.

Geographic location also seems to influence the death rate. Particularly high mortality rates have been reported in some outbreaks from the United Arab Emirates (73%) and China (80%).

Geographic differences in viral virulence have been suggested, but are unproven. The mortality rate may also be influenced by the availability of rigorous supportive treatment in area hospitals.

CHAPTER 11 RIFT VALLEY FEVER

11.1 Introduction

Rift Valley fever (RVF) is a viral zoonosis (affects primarily domestic livestock, but can be passed to humans) causing fever. It is spread by the bite of infected mosquitoes, typically the *Aedes* or *Culex* genera. The disease is caused by the RVF virus, a member of the genus *Phlebovirus* (family Bunyaviridae). The disease was first reported among livestock in Kenya around 1915, but the virus was not isolated until 1931.[RVF outbreaks occur across sub-Saharan Africa, with outbreaks occurring elsewhere infrequently, but sometimes severely. In Egypt in 1977-78, an estimated 200,000 people were infected and there were at least 594 deaths among hospitalized patients. In Kenya in 1998, the virus claimed the lives of over 400 Kenyans. In September 2000, an outbreak was confirmed in Saudi Arabia and Yemen). On 19 Oct 2011, a case of Rift Valley fever contracted in Zimbabwe was reported in a Caucasian female traveler who returned to France after a 26-day stay in Marondera, Mashonaland East Province during July and August, 2011]but later classified as 'not confirmed.

11.2 Etiology

Rift Valley fever results from infection by the Rift Valley fever virus, an RNA virus in the genus Phlebovirus (family Bunyaviridae).

11.3 Geographic Distribution

The Rift Valley fever virus is found throughout most of Africa. The disease is endemic in southern and eastern Africa, where outbreaks occur at irregular intervals. Epidemics have also been reported in Egypt, Saudi Arabia and Yemen.

11.4 Transmission

Rift Valley fever is transmitted by mosquitoes and is usually amplified in ruminant hosts. In endemic regions, cases can occur sporadically or in epidemics. The virus appears to survive in the dried eggs of Aedes mosquitoes; epidemics are associated with the hatching of these mosquitoes during years of heavy rainfall and localized flooding. In Africa, outbreaks typically occur in savannah grasslands every 5 to 15 years and in semi-arid regions every 25 to 35 years. Once it has been amplified in animals, the RVF virus can also be transmitted by other vectors, including many mosquito species and possibly other biting insects such as ticks and midges. The virus can be transmitted in utero to the fetus. It has also been found in semen and raw milk.

Humans do not seem to be infected by casual contact with live hosts, but can be infected by aerosols or direct contact with tissues during parturition, necropsy, slaughter, laboratory procedures or meat preparation for cooking.

In utero transmission to a human infant was first reported in 2006. Both animals and humans theoretically have the potential to introduce Rift Valley, fever into new areas by infecting mosquitoes.

11.5 Disinfection

Under optimal conditions, the Rift Valley fever virus remains viable in aerosols for more than an hour at 25°C (77°F). In a neutral or alkaline pH, mixed with serum or other proteins, the virus can survive for as long as four months at 4°C (40°F) and eight years below 0°C (32°F). It is quickly destroyed by pH changes in decomposing carcasses. The Rift Valley fever virus is susceptible to low pH (≤ 6.2), lipid solvents and detergents, and solutions of sodium or calcium hypochlorite with residual chlorine content greater than 5000 ppm.

11.6 Clinical Signs

Infection with the Rift Valley fever virus usually results in an asymptomatic infection or a mild to moderate, non-fatal, flu-like illness with fever and liver abnormalities

The symptoms of uncomplicated infections may include fever, headache, generalized weakness, dizziness, weight loss, myalgia and back pain. Some patients also have stiffness of the neck, photophobia and vomiting. Most people recover spontaneously within two days to a week.

Complications including hemorrhagic fever, meningoencephalitis or ocular disease occur in a small percentage of patients. Hemorrhagic fever usually develops two to four days after the initial symptoms. The symptoms may include jaundice, hematemesis, melena, a purpuric rash, petechiae and bleeding from the gums. Hemorrhagic fever frequently progresses to frank hemorrhages, shock and death.

Ocular disease and meningoencephalitis are usually seen one to three weeks after the initial symptoms. The ocular form is characterized by retinal lesions and may result in some degree of permanent visual impairment. Death is rare in cases of ocular disease or meningoencephalitis

11.7 Diagnostic Tests

The Rift Valley fever virus can be isolated from the blood, brain, liver or other tissues; in living hosts, viremia usually occurs only during the first three days of fever. The virus can be grown in numerous cell lines including baby hamster kidney cells, monkey kidney (Vero) cells, chicken embryo reticulum, and primary cultures from cattle or sheep. Hamsters, adult or suckling mice, embryonated chicken eggs or 2-day-old lambs can also be used.

Viral antigens and RNA can be detected in blood and tissue samples by various antigen detection tests and reverse transcription polymerase chain reaction (RT-PCR) assays. Enzyme-linked immunoassay (ELISA) and other serologic tests can detect specific IgM or rising titers.

11.8 Treatment

No specific treatment, other than supportive care, is available; however, ribavirin has been promising in animal studies. Interferon, immune modulators and convalescent phase plasma may also prove to be helpful. Most cases of Rift Valley fever are relatively mild, brief illnesses and may not require treatment.

11.9 Prevention

Mosquito repellents, long shirts and trousers, bednets, and other arthropod control measures should be used to prevent transmission by mosquitoes and other potential insect vectors. Outdoor activities should be avoided, if possible, during periods of peak mosquito activity.

Insecticides may be helpful. During epidemics, vaccination of susceptible animals can prevent amplification of the virus and protect people as well as animals.

Barrier precautions should be used whenever contact may occur with infectious tissues or blood from animals; recommended measures include personal protective equipment such as protective clothing, gloves and goggles.

Diagnostic tissue samples should be processed by trained staff in appropriately equipped laboratories. Universal precautions are recommended for healthcare workers who care for patients with confirmed or suspected Rift Valley fever. Barrier techniques are recommended when nursing hospitalized patients.

A human vaccine has been developed, but has limited availability. Additional vaccines are under investigation.

11.10 Morbidity and Mortality

Humans are highly susceptible to Rift Valley fever. Most cases develop in veterinarians, abattoir workers and others who work closely with blood and tissue samples from animals. During outbreaks in animals, mosquitoes may spread the virus to humans and cause epidemics. In

Egypt, approximately 200,000 human cases and 598 deaths occurred during an epidemic in 1977.

In December 2006, an outbreak of RVF in Kenya, Somalia and the United Republic of Tanzania resulted in substantial numbers of human and animal cases and deaths.

As of May 18, 2007, over 1000 human cases and 300 deaths have been reported .Most people with Rift Valley fever recover spontaneously within a week. Ocular disease is seen in approximately 0.5% to 2% of cases, and meningoencephalitis and hemorrhagic fever in less than 1%. The case fatality rate for hemorrhagic fever is approximately 50%. Deaths rarely occur in people with eye disease or meningoencephalitis, but 1% to 10% of patients with ocular disease have some permanent visual impairment. The overall case fatality rate for all patients with Rift Valley fever is less than 1%.

11.11 Infections in Animals

11.11.1 Species Affected

Rift Valley fever can affect many species of animals including sheep, cattle, goats, buffalo, camels, and monkeys, as well as gray squirrels and other rodents. The primary amplifying hosts are sheep and cattle. Viremia without severe disease may be seen in adult cats, dogs, horses and some monkeys, but severe disease can occur in newborn puppies and kittens. Rabbits, pigs, guinea pigs, chickens and hedgehogs do not become viremic.

11.11.2 Incubation Period

The incubation period can be as long as 3 days in sheep, cattle, goats and dogs. In newborn lambs, it is 12 to 36 hours. Experimental infections usually become evident after 12 hours in newborn lambs, calves, kids and puppies.

11.11.3 Clinical Signs

The clinical signs vary with the age, species and breed of the animal. In endemic regions, epidemics of Rift Valley fever can be recognized by high mortality rates in newborn animals and abortions in adults.

Rift Valley fever is usually most severe in young animals. In lambs, a biphasic fever, anorexia and lymphadenopathy may be followed by weakness and death within 36 hours. Hemorrhagic diarrhea or abdominal pain can also be seen. The youngest animals are most severely affected; in neonates, the mortality rate may reach 90% to 100%. Similar symptoms occur in young calves: fever, anorexia and depression are typical, with mortality rates of 10% to 70%.

Abortions are the most characteristic sign in adult sheep and cattle. Other symptoms that may occur in adult sheep include fever, weakness, a mucopurulent nasal discharge (sometimes bloodstained), melena, hemorrhagic or foul-smelling diarrhea, and vomiting. In adult cattle, fever, anorexia, weakness, excessive salivation, fetid diarrhea and decreased milk production have been reported.

Icterus may also be seen, particularly in cattle. Similar but milder infections occur in goats. Adult camels do not develop symptoms other than abortion, but young animals may have more severe disease. Viremia without severe disease may be seen in adult cats, dogs, horses and some monkeys, but severe disease can occur in newborn puppies and kittens.

11.11.4 Post Mortem Lesions

The most consistent lesion is hepatic necrosis; the necrosis is more extensive and severe in younger animals.

In aborted fetuses and newborn lambs, the liver may be very large, yellowish-brown to dark reddish-brown, soft and friable, with irregular patches of congestion. Multiple ray to white necrotic foci are usually present, but may only be visible microscopically. The liver lesions are usually less severe in adult animals and may consist of numerous pinpoint reddish to grayish-white necrotic foci.

Additional lesions may include jaundice, widespread cutaneous hemorrhages and fluid in the body cavities. The peripheral lymph nodes and spleen are typically enlarged and edematous, and often contain petechiae. The walls of the gallbladder are often edematous, with visible hemorrhages. A variable degree of inflammation or hemorrhagic enteritis can sometimes be found in the intestines. In lambs, numerous small hemorrhages typically occur in the abomasal mucosa, and the small intestine and abomasum may contain dark chocolate-brown contents with partially digested blood. In addition, petechial and ecchymotic hemorrhages may be seen on the surface of other internal organs. Microscopically, hepatic necrosis is the most prominent lesion.

11.11.5 Communicability

Infections in animals are typically transmitted by mosquitoes and not by direct contact; however, during parturition, necropsy or slaughter, viruses in the tissues can become aerosolized or enter the skin through abrasions. The Rift Valley fever virus has also been found in raw milk and may be present in semen.

11.11.6 Diagnostic Tests

Rift Valley fever can be diagnosed by isolation of the virus from the blood of febrile animals. The RVF virus can also be recovered from the tissues of dead animals and aborted fetuses; the liver, spleen and brain are generally used. This virus can be grown in numerous cell lines including baby hamster kidney cells, monkey kidney (Vero) cells, chicken embryo reticulum and primary cultures from cattle or sheep. Hamsters, adult or suckling mice, embryonated chicken eggs or two-day-old lambs can also be used.

Viral titers in tissues are often high, and a rapid diagnosis can sometimes be made with complement fixation, neutralization or agar gel diffusion tests on tissue suspensions. Viral antigens can also be detected by immunofluorescent staining of impression smears from the liver, spleen or brain. Enzyme immunoassays and immunodiffusion tests can identify virus in the blood. RT-PCR testing can detect viral RNA.

Commonly used serologic tests include virus neutralization, ELISA and hemagglutination inhibition tests. Immunofluorescence, complement fixation, radioimmunoassay and immunodiffusion are used less frequently. Cross-reactions with other phleboviruses can occur in serologic tests other than virus neutralization.

11.11.7 Treatment

No specific treatment, other than supportive care, is available.

11.11.8 Prevention

Vaccines are generally used to protect animals from Rift Valley fever in endemic regions. During epidemics, vaccination of susceptible animals can prevent amplification of the virus and protect people as well as animals. Attenuated and inactivated Rift Valley fever vaccines are both available. Attenuated vaccines produce better immunity; however, abortions and birth defects can occur in pregnant animals. Subunit vaccines are in development.

.Additional, less commonly used, preventative measures include vector controls, movement of stock to higher altitudes, and the confinement of stock in insect-proof stables. These control methods are often impractical, or are ineffective because they are instituted too late. The movement of animals from endemic areas to RVF-free regions can result in epidemics.

CHAPTER 12 EBOLA VIRUS

12.1 Introduction

Ebola virus (EBOV) causes extremely severe disease in humans and in nonhuman primates in the form of viral hemorrhagic fever. EBOV is a select agent, World Health Organization Risk Group 4 Pathogen (requiring Biosafety Level 4-equivalent containment), National Institutes of Health/National Institute of Allergy and Infectious Diseases Category A Priority Pathogen, Centers for Disease Control and Prevention Category A Bioterrorism Agent, and listed as a Biological Agent for Export Control by the Australia Group.

Infection typically begins with flu-like symptoms, which often progress rapidly to fatal complications of hemorrhage, fever, and hypotensive shock.

The Ebola virus was first identified in the western equatorial province of Sudan and in a nearby region of Zaire (now Democratic Republic of the Congo) in 1976 after significant epidemics in Nzara, southern Sudan and Yambuku, northern Zaire.

There are five distinct species of the Ebola virus: Bundibugyo, Côte d'Ivoire, Reston, Sudan and Zaïre. Bundibugyo, Sudan and Zaïre species have been associated with large outbreaks of Ebola haemorrhagic fever (EHF) in Africa causing death in 25-90% of all clinically ill cases, while Côte d'Ivoire and Reston have not.

The Ebola virus is transmitted by direct contact with the blood, body fluids and tissues of infected persons. Transmission of the Ebola virus has also occurred by handling sick or dead infected wild animals (chimpanzees, gorillas, monkeys, forest antelope, fruit bats). The predominant treatment is general supportive therapy.

12.2 History of Ebola

Ebola virus (abbreviated EBOV) was first described in 1976 by David Finkes. In 1976, Ebola (named after the Ebola River in Zaire) first emerged in Sudan and Zaire. The first outbreak of Ebola (Ebola-Sudan) infected over 284 people, with a mortality rate of 53%. A few months later, the second Ebola virus emerged from Yambuku, Zaire, Ebola-Zaire (EBOZ). EBOZ, with the highest mortality rate of any of the Ebola viruses (88%), infected 318 people. Despite the tremendous effort of experienced and dedicated researchers, Ebola's natural reservoir was never identified. The third strain of Ebola, Ebola Reston (EBOR), was first identified in 1989 when infected monkeys were imported into Reston, Virginia, from Mindanao in the Philippines. Fortunately, the few people who were infected with EBOR (seroconverted) never developed Ebola hemorrhagic fever (EHF). The last known strain of Ebola, Ebola Cote d'Ivoire (EBO-CI) was discovered in 1994 when a female ethologist performing a necropsy on a dead chimpanzee from the Tai Forest, Cote d'Ivoire, accidentally infected herself during the necropsy.

12.3 Classification of Ebola

The genera Ebolavirus and Marburgvirus were originally classified as the species of the now-obsolete Filovirus genus. In March 1998, the Vertebrate Virus Subcommittee proposed in the International Committee on Taxonomy of Viruses (ICTV) to change the Filovirusgenus to the Filoviridae family with two specific genera: Ebola-like viruses and Marburg-like viruses. This proposal was implemented in Washington, DC on April 2001 and in Paris on July 2002. In 2000, another proposal was made in Washington, DC, to change the "-like viruses" to "-virus" resulting in today's Ebolavirus and Marburgvirus.

12.4 Ebola species subtypes

The five characterised Ebola species are:

12.4.1 Zaire ebolavirus (ZEBOV)

Also known simply as the *Zaire virus*, ZEBOV has the highest case-fatality rate of the ebolaviruses, up to 90% in some epidemics, with an average case fatality rate of approximately 83% over 27 years. There have been more outbreaks of *Zaire ebolavirus* than of any other species. The first outbreak occurred on 26 August 1976 in Yambuku. The first recorded case was

Mabalo Lokela, a 44year-old schoolteacher. The symptoms resembled malaria, and subsequent patients received quinine. Transmission has been attributed to reuse of unsterilized needles and close personal contact.

12.4.2 Sudan ebolavirus (SEBOV)

Like the *Zaire virus*, SEBOV emerged in 1976; it was at first assumed to be identical with the Zaire species. SEBOV is believed to have broken out first among cotton factory workers in Nzara, Sudan, with the first case reported as a worker exposed to a potential natural reservoir. The virus was not found in any of the local animals and insects that were tested in response. The carrier is still unknown. The lack of barrier nursing (or "bedside isolation") facilitated the spread of the disease. The most recent outbreak occurred in May, 2004. Twenty confirmed cases were reported in Yambio County, Sudan, with five deaths resulting. The average fatality rates for SEBOV were 54% in 1976, 68% in 1979, and 53% in 2000 and 2001.

12.4.3 Reston ebolavirus (REBOV)

Discovered during an outbreak of simian hemorrhagic fever virus (SHFV) in crab-eating macaques from Hazleton Laboratories (now Covance) in 1989. Since the initial outbreak in Reston, Virginia, it has since been found in non-human primates in Pennsylvania, Texas and Siena, Italy. In each case, the affected animals had been imported from a facility in the Philippines, where the virus has also infected pigs. Despite its status as a Level4 organism and its apparent pathogenicity in monkeys, REBOV did not cause disease in exposed human laboratory workers.

12.4.4 Côte d'Ivoire ebolavirus (CIEBOV)

Also referred to as *Taï Forest ebolavirus* and by the English place name, "Ivory Coast", it was first discovered among chimpanzees from the Taï Forest in Côte d'Ivoire, Africa, in 1994. Necropsies showed blood within the heart to be brown; no obvious marks were seen on the organs; and one necropsy showed lungs filled with blood. Studies of tissues taken from the chimpanzees showed results similar to human cases during the 1976 Ebola outbreaks in Zaire and Sudan. As more dead chimpanzees were discovered, many tested positive for Ebola using molecular techniques. The source of the virus was believed to be the meat of infected Western Red Colobus monkeys, upon which the chimpanzees preyed. One of the scientists performing the necropsies on the infected chimpanzees contracted Ebola. She developed symptoms similar to those of dengue fever approximately a week after the necropsy, and was transported to Switzerland for treatment. She was discharged from the hospital after two weeks and had fully recovered six weeks after the infection.

12.4.5 Bundibugyo ebolavirus (BEBOV)

On 24 November 2007, the Uganda Ministry of Health confirmed an outbreak of Ebolavirus in the Bundibugyo District. After confirmation of samples tested by the United States National Reference Laboratories and the CDC, the World Health Organization confirmed the presence of the new species. On 20 February 2008, the Uganda Ministry officially announced the end of the epidemic in Bundibugyo, with the last infected person discharged on 8 January 2008.[11] An epidemiological study conducted by WHO and Uganda Ministry of Health scientists determined there were 116 confirmed and probable cases of the new Ebola species, and that the outbreak had a mortality rate of 34% (39 deaths). In 2012, there was an outbreak of Bundibugyo ebolavirus in a northeastern province of the Democratic Republic of the Congo. There were 15 confirmed cases and 10 fatalities.

12.5 Signs and symptoms of Ebola hemorrhagic fever

Ebola Virus Disease begins with a sudden onset of an influenza-like stage characterized by general malaise, fever with chills, arthralgia, myalgia, and chest pain. Nausea is accompanied by abdominal pain, diarrhea, and vomiting. Respiratory tract involvement is characterized by pharyngitis with sore throat, cough, dyspnea, and hiccups. The central nervous system is affected as judged by the development of severe headaches, agitation, confusion, fatigue, depression, seizures, and sometimes coma.

Cutaneous presentation may include: maculopapular rash, petechiae, purpura, ecchymoses, and hematomas (especially around needle injection sites). Development of hemorrhagic symptoms is generally indicative of a negative prognosis. However, contrary to popular belief, hemorrhage does not lead to hypovolemia and is not the cause of death (total blood loss is low except during labor). Instead, death occurs due to multiple organ dysfunction syndrome (MODS) due to fluid redistribution, hypotension, disseminated intravascular coagulation, and focal tissue necroses.

The mean incubation period, best calculated currently for EVD outbreaks due to EBOV infection, is 12.7 days (standard deviation = 4.3 days), but can be as long as 25 days

12.6 Risk factors

Between 1976 and 1998, from 30,000 mammals, birds, reptiles, amphibians, and arthropods sampled from outbreak regions, no *ebola virus* was detected apart from some genetic traces found in six rodents (*Mus setulosus* and *Praomys*) and one shrew (*Sylvisorex ollula*) collected from the Central African Republic. Traces of EBOV were detected in the carcasses of gorillas and chimpanzees during outbreaks in 2001 and 2003, which later became the source of human infections. However, the high lethality from infection in these species makes them unlikely as a natural reservoir.

Plants, arthropods, and birds have also been considered as possible reservoirs; however, bats are considered the most likely candidate. Bats were known to reside in the cotton factory in which the index cases for the 1976 and 1979 outbreaks were employed, and they have also been implicated in Marburg virus infections in 1975 and 1980. Of 24 plant species and 19 vertebrate species experimentally inoculated with EBOV, only bats became infected. The absence of clinical signs in these bats is characteristic of a reservoir species. In a 2002–2003 survey of 1,030 animals which included 679 bats from Gabon and the Republic of the Congo, 13 fruit bats were found to contain EBOV RNA fragments.[1] As of 2005, three types of fruit bats (*Hypsignathus monstrosus*, *Epomops franqueti*, and *Myonycteris torquata*) have been identified as being in contact with EBOV. They are now suspected to represent the EBOV reservoir hosts.

The existence of integrated genes of filoviruses in some genomes of small rodents, insectivorous bats, shrews, tenrecs, and marsupials indicates a history of infection with filoviruses in these groups as well. However, it has to be stressed that infectious ebolaviruses have not yet been isolated from any nonhuman animal.

Bats drop partially eaten fruits and pulp, then terrestrial mammals such as gorillas and duikers feed on these fallen fruits. This chain of events forms a possible indirect means of transmission from the natural host to animal populations, which have led to research towards viral shedding in the saliva of bats. Fruit production, animal behavior, and other factors vary at different times and places which may trigger outbreaks among animal populations. Transmission between natural reservoirs and humans are rare, and outbreaks are usually traceable to a single index case where an individual has handled the carcass of gorilla, chimpanzee, or duiker The virus then spreads person-to-person, especially within families, hospitals, and during some mortuary rituals where contact among individuals becomes more likely

The virus has been confirmed to be transmitted through body fluids. Transmission through oral exposure and through conjunctiva exposure is likely and has been confirmed in non-human primates. Filoviruses are not naturally transmitted by aerosol. They are, however, highly infectious as breathable 0.8–1.2 micrometre droplets in laboratory conditions; because of this potential route of infection, these viruses have been classified as Category A biological weapons.

All epidemics of Ebola have occurred in sub-optimal hospital conditions, where practices of basic hygiene and sanitation are often either luxuries or unknown to caretakers and where disposable needles and autoclaves are unavailable or too expensive. In modern hospitals with disposable needles and knowledge of basic hygiene and barrier nursing techniques, Ebola has never spread on a large scale. In isolated settings such as a quarantined hospital or a remote village, most victims are infected shortly after the first case of infection is present. The quick onset of symptoms from the time the disease becomes contagious in an individual makes it easy to identify sick individuals and limits an individual's ability to spread the disease by traveling. Because bodies of the deceased are still infectious, some doctors had to take measures to properly dispose of dead bodies in a safe manner despite local traditional burial rituals

12.7 Virology

12.7.1 Genome

Like all mononegaviruses, ebolavirions contain linear nonsegmented, single-stranded, non-infectious RNA genomes of negative polarity that possesses inverse-complementary 3' and 5' termini, do not possess a 5' cap, are not polyadenylated, and are not covalently linked to a protein Ebolavirus genomes are approximately 19 kilobase pairs long and contain seven genes in the order 3'-UTR-NP-VP35-VP40-GP-VP30-VP24-L-5'-UTR. The genomes of the five different ebolaviruses (BDBV, EBOV, RESTV, SUDV, and TAFV) differ in sequence and the number and location of gene overlaps.

12.7.2. Structure

Like all filoviruses, ebolavirions are filamentous particles that may appear in the shape of a shepherd's crook or in the shape of a "U" or a "6", and they may be coiled, toroid, or branched

Ebolavirions are generally 80 nm in width, but vary somewhat in length. In general, the median particle length of ebolaviruses ranges from 974–1,086 nm (in contrast to marburgvirions, whose median particle length was measured to be 795–828 nm), but particles as long as 14,000 nm have been detected in tissue culture. Ebolavirions consist of seven structural proteins. At the center is the helical ribonucleocapsid, which consists of the genomic RNA wrapped around a polymer of nucleoproteins (NP). Associated with the ribonucleoprotein is the RNA-dependent RNA polymerase (L) with the polymerase cofactor (VP35) and a transcription activator (VP30). The ribonucleoprotein is embedded in a matrix, formed by the major (VP40) and minor (VP24) matrix proteins. These particles are surrounded by a lipid membrane derived from the host cell membrane. The membrane anchors a glycoprotein ($GP_{1,2}$) that projects 7 to 10 nm spikes away from its surface. While nearly identical to marburgvirions in structure, ebolavirions are antigenically distinct.

12.7.3 Entry

Niemann–Pick C1 (NPC1) appears to be essential for Ebola infection. Two independent studies reported in the same issue of Nature showed that Ebola virus cell entry and replication requires the cholesterol transporter protein NPC1. When cells from Niemann Pick Type C1 patients (who have a mutated form of NPC1) were exposed to Ebola virus in the laboratory, the cells survived and appeared immune to the virus, further indicating that Ebola relies on NPC1 to enter cells. This might imply that genetic mutations in the NPC1 gene in humans could make some people resistant to one of the deadliest known viruses affecting humans. The same studies described similar results with Ebola's cousin in the filovirus group, Marburg virus, showing that it too needs NPC1 to enter cells. Furthemore, NPC1 was shown to be critical to filovirus entry because it mediates infection by binding directly to the viral envelope glycoprotein. A later study confirmed the findings that NPC1 is a critical filovirus receptor that mediates infection by binding directly to the viral envelope glycoprotein and that the second lysosomal domain of NPC1 mediates this binding

In one of the original studies, a small molecule was shown to inhibit Ebola virus infection by preventing the virus glycoprotein from binding to NPC1. In the other study, mice that were heterozygous for NPC1 were shown to be protected from lethal challenge with mouse adapted Ebola virus . Together, these studies suggest NPC1 may be potential therapeutic target for an Ebola anti-viral drug.

12.7.4. Replication

The ebolavirus life cycle begins with virion attachment to specific cell-surface receptors, followed by fusion of the virion envelope with cellular membranes and the concomitant release of the virus nucleocapsid into the cytosol. The viral RNA polymerase, encoded by the L gene, partially uncoats the nucleocapsid and transcribes the genes into positive-stranded mRNAs, which are then translated into structural and nonstructural proteins. Ebolavirus RNA polymerase (L) binds to a single promoter located at the 3' end of the genome. Transcription either terminates after a gene or continues to the next gene downstream. This means that genes close to the 3' end of the genome are transcribed in the greatest abundance, whereas those toward the 5' end are least likely to be transcribed. The gene order is therefore a simple but effective form of transcriptional regulation. The most abundant protein produced is the nucleoprotein, whose concentration in the cell determines when L switches from gene transcription to genome replication. Replication results in full-length, positive-stranded antigenomes that are, in turn, transcribed into negative-stranded virus progeny genome copy. Newly synthesized structural proteins and genomes self-assemble and accumulate near the inside of the cell membrane. Virions bud off from the cell, gaining their envelopes from the cellular membrane they bud from. The mature progeny particles then infect other cells to repeat the cycle

12.8 Pathophysiology

Endothelial cells, mononuclear phagocytes, and hepatocytes are the main targets of infection. After infection, in a secreted glycoprotein (sGP) the Ebola virus glycoprotein (GP) is synthesized. Ebola replication overwhelms protein synthesis of infected cells and host immune defenses. The GP forms a trimeric complex, which binds the virus to the endothelial cells lining the interior surface of blood vessels. The sGP forms a dimeric protein which interferes with the signaling of neutrophils, a type of white blood cell, which allows the virus to evade the immune system by inhibiting early steps of neutrophil activation. These white blood cells also serve as carriers to transport the virus throughout the entire body to places such as the lymph nodes, liver, lungs, and spleen. The presence of viral particles and cell damage resulting from budding causes the release of cytokines (specifically TNF-α, IL-6, IL-8, etc.), which are the signaling molecules for fever and inflammation. The cytopathic effect, from infection in the endothelial cells, results in a loss of vascular integrity. This loss in vascular integrity is furthered with synthesis of GP, which reduces specific integrins responsible for cell adhesion to the inter-cellular structure, and damage to the liver, which leads to coagulopathy

12.9 Diagnosis

VD is clinically indistinguishable from Marburg virus disease (MVD), and it can also easily be confused with many other diseases prevalent in Equatorial Africa, such as other viral hemorrhagic fevers, falciparum malaria, typhoid fever, shigellosis, rickettsial diseases such as typhus, cholera, gram-negative septicemia, borreliosis such as relapsing fever or EHEC enteritis. Other infectious diseases that ought to be included in the differential diagnosis include leptospirosis, scrubtyphus, plague, candidiasis, histoplasmosis, trypanosomiasis, visceral leishmaniasis, hemorrhagic smallpox, measles, and fulminant viral hepatitis. Non-infectious diseases that can be confused with EVD are acute promyelocytic leukemia, hemolytic uremic syndrome, snake envenomation, clotting factor deficiencies/platelet disorders, thrombotic thrombocytopenic purpura, hereditary hemorrhagic telangiectasia, Kawasaki disease, and even warfarin intoxication.

12.10 Prevention

Ebola viruses are highly infectious as well as contagious.

As an outbreak of ebola progresses, bodily fluids from diarrhea, vomiting, and bleeding represent a hazard. Due to lack of proper equipment and hygienic practices, large-scale epidemics occur mostly in poor, isolated areas without modern hospitals or well-educated medical staff. Many areas where the infectious reservoir exists have just these characteristics. In such environments, all that can be done is to immediately cease all needle-sharing or use without adequate sterilization procedures, isolate patients, and observe strict barrier nursing procedures with the use of a medical-rated disposable face mask, gloves, goggles, and a gown at all times, strictly enforced for all medical personnel and visitors. The aim of all of these techniques is to avoid any person's contact with the blood or secretions of any patient, including those who are deceased.

Vaccines have successfully protected nonhuman primates; however, the six months needed to complete immunization made it impractical in an epidemic. To resolve this, in 2003, a vaccine using an adenoviral (ADV) vector carrying the Ebola spike protein was tested on crab-eating macaques. The monkeys were challenged with the virus 28 days later, and remained resistant. In 2005, a vaccine based on attenuated recombinant vesicular stomatitis virus (VSV) vector carrying either the Ebola glycoprotein or Marburg glycoprotein successfully protected nonhuman primates, opening clinical trials in humans. By October, the study completed the first human trial; giving three vaccinations over three months showing capability of safely inducing an immune response. Individuals were followed for a year, and, in 2006, a study testing a faster-acting, single-shot vaccine began. This study was completed in 2008. The next step is to try the vaccine on a strain of Ebola that is closer to the one that infects humans. There are currently no Food and Drug Administration-approved vaccines for the prevention of EVD. Many candidate vaccines have been developed and tested in various animal models. Of those, the most promising ones are DNA vaccines or are based on adenoviruses, vesicular stomatitis Indiana virus (VSIV) or filovirus-like particles (VLPs) as all of these candidates could protect nonhuman primates from ebolavirus-induced disease. DNA vaccines, adenovirus-based vaccines, and VSIV-based vaccines have entered clinical trials.

Contrary to popular belief, ebolaviruses are not transmitted by aerosol during natural EVD outbreaks. Due to the absence of an approved vaccine, prevention of EVD therefore relies predominantly on behavior modification, proper personal protective equipment, and sterilization/disinfection.

On 6 December 2011, the development of a successful vaccine against Ebola for mice was reported. Unlike the predecessors, it can be freeze-dried and thus stored for long periods in wait for an outbreak. The research is reported in *Proceedings of National Academy of Sciences*

According to the authors (Phoolcharoen et al.) of the above article they have evaluated the immunogenicity an efficacy of an EBOV vaccine candidate in which the viral surface glycoprotein is biomanufactured as a fusion to a monoclonal antibody that recognizes an epitope in glycoprotein, resulting in the production of Ebola immune complexes (EICs).

Although antigen– antibody immune complexes are known to be efficiently processed and presented to immune effector cells, authors found that codelivery of the EIC with Toll-like receptor agonists elicited a more robust antibody response in mice than did EIC alone. Among the compounds tested by authors , polyinosinic:polycytidylic acid (PIC, a Toll-like receptor 3 agonist) was highly effective as an adjuvant agent.

After vaccinating mice with EIC plus PIC, 80% of the animals were protected against a lethal challenge with live EBOV (30,000 LD50 of mouse adapted virus). Surviving animals showed a mixed Th1/Th2 response to the antigen, suggesting this may be important for protection. Survival after vaccination with EIC plus PIC was statistically equivalent to that achieved with an alternative viral vector vaccine candidate reported in the literature. Because nonreplicating subunit vaccines offer the possibility of formulation for cost effective, long-term storage in bio threat reduction repositories, EIC is an attractive option for public health defense measures.

12.12 Recent research

In late 2012, Canadian scientists discovered that the deadliest form of the virus could be transmitted by air between species. They managed to prove that the virus was transmitted from pigs to monkeys without any direct contact between them, leading to fears that airborne transmission could be contributing to the wider spread of the disease in parts of Africa. Evidence was also found that pigs might be one of the reservoir hosts for the virus; the fruit bat has long been considered as the reservoir

CHAPTER 13 MARBURG VIRUS

13,1 Introduction

Marburg virus is a hemorrhagic fever virus first noticed and described during small epidemics in the German cities Marburg and Frankfurt and the Yugoslavian capital Belgrade in the 1960s. Workers were accidentally exposed to tissues of infected grivets (*Chlorocebus aethiops*) at the city's former main industrial plant, the Behringwerke, then part of Hoechst, and today of CSL Behring. During these outbreaks, 31 people became infected and seven of them died. Marburg virus (MARV) causes severe disease in humans and nonhuman primates in the form of viral hemorrhagic fever. MARV is a Select Agent, WHO Risk Group 4 Pathogen (requiring biosafety level 4-equivalent containment), NIH/National Institute of Allergy and Infectious Diseases Category A Priority Pathogen, Centers for Disease Control and Prevention Category A Bioterrorism Agent and is listed as a biological agent for export control by the Australia Group.

Marburgvirus (the gentle sister) affects humans somewhat like nuclear radiation, damaging virtually all of the tissues in their bodies. It attacks with particular ferocity the internal organs, connective tissue, intestines, and skin. In Germany, all the survivors lost their hair-they went bald or partly bald. Their hair died at the roots and fellout in clumps, as if they had received radiation burns. Hemorrhage occurred from all orifices of the body. During the survivors' recovery period, the skin peeled off their faces, hands, feet, and genitals. Some of the men suffered from blown up, semirotten testicles. One of the worst cases of this appeared in a morgue attendant who had handled Marburg-infected bodies. The virus also lingered in the fluid inside the eyeballs of some victims for many months

12.2 History

The year was 1967. Several laboratory workers, all from the same lab in Marburg, Germany, were hospitalized with a severe and strange disease. The physicians on staff realized the workers were all suffering from the same ailment, with symptoms that included fever, diarrhea, vomiting, massive bleeding from many different organs, shock, and eventually circulatory system collapse. An investigation began in an attempt to uncover the source of the outbreak. This led to the identification of the source of the virus in Germany: a species of African green monkeys, imported from Uganda, which were being used by the scientists for polio vaccine research. The virus was isolated, and found to exhibit a unique morphology, leading to the designation of a new group: the Filoviridae In that outbreak, a total of 31 human cases were observed, and the disease presented with a 23% mortality rate (7 deaths occurred out of 31 total infections).

After this episode, the virus went back into hiding for almost a decade, not surfacing again until 1975 in South Africa. The origin of this outbreak is unknown, although based on epidemiological studies it is assumed that the index case (the first person known to have been infected), a young man hitchhiking through Africa, acquired the disease in Zimbabwe, and then infected 2 others in South Africa when he arrived there. Only the index case died from the disease; the secondary cases (those infected due to contact with the index case) survived.

Marburg again disappeared until one case was reported in Kenya in 1980, and again in 1987 in the same area. In the first outbreak, again only the index case died, while a second patient survived. Only one infection was noted in the 1987 outbreak, resulting in death of the patient. Both of these Kenyan outbreaks occurred in the vicinity of a volcano named Mount Elgon, and there is evidence both index cases had spent time in a cave inside the mountain. This has led to (unconfirmed) speculation that bats may be a reservoir for filoviruses;

Between 1987 and 1998, the only cases of Marburg were due to laboratory accidents, both in the former Soviet Union. One of these cases was fatal. However, in 1998, the largest natural outbreak of Marburg virus disease to date began in northeastern Democratic Republic of the Congo (DRC). This time, the focus of the outbreak was a town called Durba (population 16,000). A large number of men in this region work for the Kilo Moto Mining Company, which runs a number of illegal gold mines in the area. Working in this area is precarious; civil war broke out in 1996, and the socio-economic situation has deteriorated since then. Infectious diseases of all types are common, as vaccinations and medication are in short supply. The Marburg outbreak is thought to have started in November of 1998, although it was not reported to any international agencies until late April of 1999, following the death of the chief medical officer in the area from the disease. At this time, local officials contacted the Medecins sans Frontieres (Doctors without Borders) in Belgium regarding the ongoing hemorrhagic fever epidemic. Both this group and their sister group in Holland sent officers to investigate and to stem the spread of the epidemic.

Patient samples were immediately sent to the National Institute of Virology in Johannesburg, South Africa; a diagnosis of Marburg virus as the cause of the illness was made on May 6th. Barrier nursing procedures were instituted, and isolation wards were fashioned at the hospital. Over the course of the epidemic, 149 cases with 123 deaths were recorded (83% fatality rate). Miners were found to be at a significantly higher risk of contracting Marburg than the general population of this area, suggesting they may be more frequently exposed to the natural reservoir of Marburg virus.

This outbreak was eclipsed beginning in 2004, when Marburg hit Angola in the fall. Due to civil war in the country and a non-existent public health infrastructure, the outbreak wasn't identified until almost 6 months later, in March of 2005. 252 people were infected, with 227 deaths (90% fatality rate). The outbreak was declared over in November of 2005, after no additional cases had been reported since summer.

This brings us to present-day Uganda, where another Marburg outbreak is ongoing, again associated with miners. So far, this one has been minor, and seems to be contained; meanwhile, back in the DRC, there comes a report of another outbreak of hemorrhagic fever. It's not known at the time what the causative agent is, but a hundred deaths have already been reported:

12.3 Classification

Marburg virus disease (MVD) is the official name listed in the World Health Organization's International Statistical Classification of Diseases and Related Health Problems 10 (ICD-10) for the human disease caused by any of the two marburgviruses Marburg virus(MARV) and Ravn virus (RAVV). In the scientific literature, Marburg hemorrhagic fever (MHF) is often used as an unofficial alternative name for the same disease. Both disease names are derived from the German city Marburg, where MARV was first discovered.

12.4 Causes

MVD is caused by two viruses classified in the genus Marburgvirus, family Filoviridae, order Mononegavirales: Marburg virus (MARV) and Ravn virus (RAVV).

Marburgviruses are endemic in arid woodlands of equatorial Africa. Most marburgvirus infections were repeatedly associated with people visiting natural caves or working in mines. In

2009, the successful isolation of infectious MARV and RAVV was reported from healthy Egyptian rousettes (Rousettus aegyptiacus) caught in caves.

This isolation strongly suggests that Old World fruit bats are involved in the natural maintenance of marburgviruses and that visiting bat-infested caves is a risk factor for acquiring marburgvirus infections. Further studies are necessary to establish whether Egyptian rousettes are the actual hosts of MARV and RAVV or whether they get infected via contact with another animal and therefore serve only as intermediate hosts. Another risk factor is contact with nonhuman primates, although only one outbreak of MVD (in 1967) was due to contact with infected monkeys.[1] Finally, a major risk factor for acquiring marburgvirus infection is occupational exposure, i.e. treating patients with MVD without proper personal protective equipment.

Contrary to Ebola virus disease (EVD), which has been associated with heavy rains after long periods of dry weather, triggering factors for spillover of marburgviruses into the human population have not yet been described.

12.5 Epidemiology

Epidemiology of Marburg virus infections

Year	Virus	Geographic location	Human cases/deaths (case-fatality rate)
1967	MARV	Marburg and Frankfurt, West Germany; and Belgrade, Yugoslavia	31/7 (23%)
1975	MARV	Rhodesia and Johannesburg, South Africa	3/1 (33%)
1980	MARV	Kenya	2/1 (50%)
1987	RAVV	Kenya	1/1 (100%)[
1988	MARV	Koltsovo, Soviet Union	1/1 (100%)

			[laboratory accident]
1990	MARV	Koltsovo, Soviet Union	1/0 (0%) [laboratory accident]
1998–2000	MARV + RAVV	Durba and Watsa, Democratic Republic of the Congo	154/128 (83%
2004–2005	MARV	Angola	252/227 (90%)
2007	MARV + RAVV	Uganda	4/1 (25%
2008	MARV	Uganda, Netherlands, Colorado	2/1 (50%)

MVD was first documented in 1967, when 31 people became ill in the German towns of Marburg and Frankfurt am Main, and in Belgrade, Yugoslavia. The outbreak involved 25 primary MARV infections and seven deaths, and six nonlethal secondary cases. The outbreak was traced to infected grivets (species *Chlorocebus aethiops*) imported from an undisclosed location in Uganda and used in developing poliomyelitis vaccines. The monkeys were received by Behringwerke, a Marburg company founded by the first winner of the Nobel Prize in Medicine, Emil von Behring. The company, which at the time was owned by Hoechst, was originally set up to develop sera against tetanus and diphtheria. Primary infections occurred in Behringwerke laboratory staff while working with grivet tissues or tissue cultures without adequate personal protective equipment. Secondary cases involved two physicians, a nurse, a post-mortem attendant, and the wife of a veterinarian. All secondary cases had direct contact, usually involving blood, with a primary case. Both physicians became infected through accidental skin pricks when drawing blood from patients.

12.6 Virology

12.6.1 Genome
Like all mononegaviruses, marburgvirions contain non-infectious, linear nonsegmented, single-stranded RNA genomes of negative polarity that possesses inverse-complementary 3' and 5' termini, do not possess a 5' cap, are not polyadenylated, and are not covalently linked to a protein. Marburgvirus genomes are approximately 19 kb long and contain seven genes in the order 3'-UTR-*NP-VP35-VP40-GP-VP30-VP24-L*-5'-UTR The genomes of the two different marburgviruses (MARV and RAVV) differ in sequence.

12.6.2 *Structure*
Like all filoviruses, marburgvirions are filamentous particles that may appear in the shape of a shepherd's crook or in the shape of a "U" or a "6", and they may be coiled, toroid, or branched. Marburgvirions are generally 80 nm in width, but vary somewhat in length. In general, the median particle length of marburgviruses ranges from 795–828 nm (in contrast to ebolavirions, whose median particle length was measured to be 974–1,086 nm), but particles as long as 14,000 nm have been detected in tissue culture. Marburgvirions consist of seven structural proteins. At the center is the helical ribonucleocapsid, which consists of the genomic RNA wrapped around a polymer of nucleoproteins (NP). Associated with the ribonucleoprotein is the RNA-dependent RNA polymerase (L) with the polymerase cofactor (VP35) and a transcription activator (VP30). The ribonucleoprotein is embedded in a matrix, formed by the major (VP40) and minor (VP24) matrix proteins. These particles are surrounded by a lipid membrane derived from the host cell membrane. The membrane anchors a glycoprotein ($GP_{1,2}$) that projects 7 to 10 nm spikes away from its surface. While nearly identical to ebolavirions in structure, marburgvirions are antigenically distinct.

12.6.3 Replication
The marburgvirus life cycle begins with virion attachment to specific cell-surface receptors, followed by fusion of the virion envelope with cellular membranes and the concomitant release of the virus nucleocapsid into the cytosol. The virus RdRp partially uncoats the nucleocapsid and transcribes the genes into positive-stranded mRNAs, which are then translated into structural and nonstructural proteins. Marburgvirus L binds to a single promoter located at the 3' end of the genome. Transcription either terminates after a gene or continues to the next gene downstream. This means that genes close to the 3' end of the genome are transcribed in the greatest abundance, whereas those toward the 5' end are least likely to be transcribed. The gene order is

therefore a simple but effective form of transcriptional regulation. The most abundant protein produced is the nucleoprotein, whose concentration in the cell determines when L switches from gene transcription to genome replication. Replication results in full-length, positive-stranded antigenomes that are in turn transcribed into negative-stranded virus progeny genome copies. Newly synthesized structural proteins and genomes self-assemble and accumulate near the inside of the cell membrane. Virions bud off from the cell, gaining their envelopes from the cellular membrane they bud from. The mature progeny particles then infect other cells to repeat the cycle

There are currently no Food and Drug Administration-approved vaccines for the prevention of MVD. Many candidate vaccines have been developed and tested in various animal models Of those, the most promising ones are DNA vaccines or based on Venezuelan equine encephalitis virus replicons, vesicular stomatitis Indiana virus (VSIV) or filovirus-like particles (VLPs) as all of these candidates could protect nonhuman primates from marburgvirus-induced disease. DNA vaccines have entered clinical trials. Marburgviruses are highly infectious, but not very contagious. Importantly, and contrary to popular belief, marburgviruses do not get transmitted by aerosol during natural MVD outbreaks. Due to the absence of an approved vaccine, prevention of MVD therefore relies predominantly on behavior modification, proper personal protective equipment, and sterilization/disinfection.

The natural maintenance hosts of marburg viruses remain to be identified unequivocally. However, the isolation of both MARV and RAVV from bats and the association of several MVD outbreaks with bat-infested mines or caves strongly suggests that bats are involved in marburg virus transmission to humans. Avoidance of contact with bats and abstaining from visits to caves is highly recommended, but may not be possible for those working in mines or people dependent on bats as a food source.

Since marburgviruses are not spreading via aerosol, the most straightforward prevention method during MVD outbreaks is to avoid direct (skin-to-skin) contact with patients, their excretions and body fluids, or possibly contaminated materials and utensils. Patients ought to be isolated but still have the right to be visited by family members. Medical staff should be trained and apply strict barrier nursing techniques (disposable face mask, gloves, goggles, and a gown at all times). Traditional burial rituals, especially those requiring embalming of bodies, ought to be discouraged or modified, ideally with the help of local traditional healers.

Marburgviruses are World Health Organization Risk Group 4 Pathogens, requiring Biosafety Level 4-equivalent containment. Laboratory researchers have to be properly trained in BSL-4 practices and wear proper personal protective equipment.

12.7 Treatment

There is currently no effective marburgvirus-specific therapy for MVD. Treatment is primarily supportive in nature and includes minimizing invasive procedures, balancing fluids and electrolytes to counter dehydration, administration of anticoagulants early in infection to prevent or control disseminated intravascular coagulation, administration of procoagulants late in infection to control hemorrhaging, maintaining oxygen levels, pain management, and administration of antibiotics or antimycotics to treat secondary infections. Experimentally, recombinant vesicular stomatitis Indiana virus (VSIV) expressing the glycoprotein of MARV has been used successfully in nonhuman primate models as post-exposure prophylaxis. Novel, very promising, experimental therapeutic regimens rely on antisense technology: phosphorodiamidate morpholino oligomers (PMOs) targeting the MARV genome could prevent disease in nonhuman primates

CHAPTER 14 DENGUE

14.1 Introduction and History

Dengue fever, also known as breakbone fever, is an infectious tropical disease caused by the dengue virus. Symptoms include fever, headache, muscle and joint pains, and a characteristic skin rash that is similar to measles. In a small proportion of cases the disease develops into the life-threatening dengue hemorrhagic fever, resulting in bleeding, low levels of blood platelets and blood plasma leakage, or into dengue shock syndrome, where dangerously low blood pressure occurs.

Dengue is transmitted by several species of mosquito within the genus *Aedes*, principally *A. aegypti*. The virus has four different types; infection with one type usually gives lifelong immunity to that type, but only short-term immunity to the others. Subsequent infection with a different type increases the risk of severe complications. As there is no commercially available vaccine, prevention is sought by reducing the habitat and the number of mosquitoes and limiting exposure to bites.

Treatment of acute dengue is supportive, using either oral or intravenous rehydration for mild or moderate disease, and intravenous fluids and blood transfusion for more severe cases. The incidence of dengue fever has increased dramatically since the 1960s, with around 50–100 million people infected yearly. Early descriptions of the condition date from 1779, and its viral cause and the transmission were elucidated in the early 20th century. Dengue has become a global problem since the Second World War and is endemic in more than 110 countries. Apart from eliminating the mosquitoes, work is ongoing on a vaccine, as well as medication targeted directly at the virus.

14.2 Sign and symptoms

Typically, people infected with dengue virus are asymptomatic (80%) or only have mild symptoms such as an uncomplicated fever. Others have more severe illness (5%), and in a small proportion it is life-threatening. The incubation period (time between exposure and onset of symptoms) ranges from 3–14 days, but most often it is 4–7 days. Therefore, travelers returning from endemic areas are unlikely to have dengue if fever or other symptoms start more than 14 days after arriving home.[1] Children often experience symptoms similar to those of the common cold and gastroenteritis (vomiting and diarrhea) and have a greater risk of severe complications, though initial symptoms are generally mild but include high fever

The characteristic symptoms of dengue are sudden-onset fever, headache (typically located behind the eyes), muscle and joint pains, and a rash. The alternative name for dengue, "breakbone fever", comes from the associated muscle and joint pains. The course of infection is divided into three phases: febrile, critical, and recovery.

The febrile phase involves high fever, potentially over 40 °C (104 °F), and is associated with generalized pain and a headache; this usually lasts two to seven days. Vomiting may also occur. A rash occurs in 50–80% of those with symptoms in the first or second day of symptoms as flushed skin, or later in the course of illness (days 4–7), as a measles-like rash. petechiae (small red spots that do not disappear when the skin is pressed, which are caused by broken capillaries) can appear at this point as may some mild bleeding from the mucous membranes of the mouth and nose. The fever itself is classically biphasic in nature, breaking and then returning for one or two days, although there is wide variation in how often this pattern actually happens.

In some people, the disease proceeds to a critical phase around the time fever resolves and typically lasts one to two days. During this phase there may be significant fluid accumulation in the chest and abdominal cavity due to increased capillary permeability and leakage. This leads to depletion of fluid from the circulation and decreased blood supply to vital organs. During this phase, organ dysfunction and severe bleeding, typically from the gastrointestinal tract, may occur. Shock (dengue shock syndrome) and hemorrhage (dengue hemorrhagic fever) occur in less than 5% of all cases of dengue, however those who have previously been infected with other serotypes of dengue virus ("secondary infection") are at an increased risk. This critical phase, while rare, occurs relatively more commonly in children and young adults. The recovery phase occurs next, with resorption of the leaked fluid into the bloodstream. This usually lasts two to three days. The improvement is often striking, and can be accompanied with severe itching and a slow heart rate. Another rash may occur with either a maculopapular or a vasculitic appearance, which is followed by peeling of the skin. During this stage, a fluid overload state may occur; if it affects the brain, it may cause a reduced level of consciousness or seizures. A feeling of fatigue may last for weeks in adults.

Dengue can occasionally affect several other body systems, either in isolation or along with the classic dengue symptoms. A decreased level of consciousness occurs in 0.5–6% of severe cases, which is attributable either to infection of the brain by the virus or indirectly as a result of impairment of vital organs, for example, the liver.

Other neurological disorders have been reported in the context of dengue, such as transverse myelitis and Guillain-Barré syndrome. Infection of the heart and acute liver failure are among the rarer complications

14.3 Virology

Dengue fever virus (DENV) is an RNA virus of the family *Flaviviridae*; genus *Flavivirus*. Other members of the same genus include yellow fever virus, West Nile virus, St. Louis encephalitis virus, Japanese encephalitis virus, tick-borne encephalitis virus, Kyasanur forest disease virus, and Omsk hemorrhagic fever virus. Most are transmitted by arthropods (mosquitoes or ticks), and are therefore also referred to as arboviruses (*ar*thropod-*bo*rne viruses).

The dengue virus genome (genetic material) contains about 11,000 nucleotide bases, which code for the three different types of protein molecules (C, prM and E) that form the virus particle and seven other types of protein molecules (NS1, NS2a, NS2b, NS3, NS4a, NS4b, NS5) that are only found in infected host cells and are required for replication of the virus. There are four strains of the virus, which are called serotypes, and these are referred to as DENV-1, DENV-2, DENV-3 and DENV-4. The distinctions between the serotypes is based on the their antigenicity

14.4 Transmission

Dengue virus is primarily transmitted by Aedes mosquitoes, particularly A. aegypti. These mosquitoes usually live between the latitudes of 35° North and 35° South below an elevation of 1,000 metres (3,300 ft). They typically bite during the day, particularly in the early morning and in the evening, but they are able to bite and thus spread infection at any time of day all during the year. Other Aedesspecies that transmit the disease include A. albopictus, A. polynesiensis and A. scutellaris. Humans are the primary host of the virus, but it also circulates in nonhuman primates. An infection can be acquired via a single bite. A female mosquito that takes a blood meal from a person infected with dengue fever, during the initial 2–10 day febrile period, becomes itself infected with the virus in the cells lining its gut. About 8–10 days later, the virus spreads to other tissues including the mosquito's salivary glands and is subsequently released into its saliva. The virus seems to have no detrimental effect on the mosquito, which remains infected for life. Aedes aegypti prefers to lay its eggs in artificial water containers, to live in close proximity to humans, and to feed on people rather than other vertebrates.

Dengue can also be transmitted via infected blood products and through organ donation. In countries such as Singapore, where dengue is endemic, the risk is estimated to be between 1.6 and 6 per 10,000 transfusions. Vertical transmission (from mother to child) during pregnancy or at birth has been reported. Other person-to-person modes of transmission have also been reported, but are very unusualThe genetic variation in dengue viruses is region specific, suggestive that establishment into new territories is relatively infrequent, despite dengue emerging in new regions in recent decades.

14.5 Predisposition

Severe disease is more common in babies and young children, and in contrast to many other infections it is more common in children that are relatively well nourished. Other risk factors for severe disease include female sex, high body mass index, and viral load. While each serotype can cause the full spectrum of disease, virus strain is a risk factor. Infection with one serotype is thought to produce lifelong immunity to that type, but only short term protection against the other three. The risk of severe disease from secondary infection increases if someone previously exposed to serotype DENV-1 contracts serotype DENV-2 or DENV-3, or if someone previously exposed to DENV-3 acquires DENV-2. Dengue can be life-threatening in people with chronic diseases such as diabetes and asthma. Polymorphisms (normal variations) in particular genes have been linked with an increased risk of severe dengue complications. Examples include the genes coding for the proteins known as TNFα, mannan-binding lectin, CTLA4, TGFβ, DC-SIGN, PLCE1, and particular forms of human leukocyte antigen from gene variations of HLA-B. A common genetic abnormality in Africans, known as glucose-6-phosphate dehydrogenase deficiency, appears to increase the risk. Polymorphisms in the genes for the vitamin D receptor and FcγR seem to offer protection against severe disease in secondary dengue infection.

14.6 Mechanism

When a mosquito carrying dengue virus bites a person, the virus enters the skin together with the mosquito's saliva. It binds to and enters white blood cells, and reproduces inside the cells while they move throughout the body. The white blood cells respond by producing a number of signaling proteins, such as cytokines and interferons, which are responsible for many of the symptoms, such as the fever, the flu-like symptoms and the severe pains. In severe infection, the virus production inside the body is greatly increased, and many more organs (such as the liver and the bone marrow) can be affected. Fluid from the bloodstream leaks through the wall of small blood vessels into body cavities due to endothelial dysfunction. As a result, less blood circulates in the blood vessels, and the blood pressure becomes so low that it cannot supply sufficient blood to vital organs. Furthermore, dysfunction of the bone marrow due to infection of the stromal cells leads to reduced numbers of platelets, which are necessary for effective blood clotting; this increases the risk of bleeding, the other major complication of dengue fever.

14.7 Viral replication

Once inside the skin, dengue virus binds to Langerhans cells (a population of dendritic cells in the skin that identifies pathogens). The virus enters the cells through binding between viral proteins and membrane proteins on the Langerhans cell, specifically the C-type lectins called DC-SIGN, mannose receptor and CLEC5A. DC-SIGN, a non-specific receptor for foreign material on dendritic cells, seems to be the main point of entry. The dendritic cell moves to the nearest lymph node. Meanwhile, the virus genome is translated in membrane-bound vesicles on the cell's endoplasmic reticulum, where the cell's protein synthesis apparatus produces new viral proteins that replicate the viral RNA and begin to form viral particles. Immature virus particles are transported to the Golgi apparatus, the part of the cell where some of the proteins receive necessary sugar chains (glycoproteins). The now mature new viruses bud on the surface of the infected cell and are released by exocytosis. They are then able to enter other white blood cells, such as monocytes and macrophages.

The initial reaction of infected cells is to produce interferon, a cytokine that raises a number of defenses against viral infection through the innate immune system by augmenting the production of a large group of proteins mediated by the JAK-STAT pathway. Some serotypes of dengue virus appear to have mechanisms to slow down this process. Interferon also activates the adaptive immune system, which leads to the generation of antibodies against the virus as well as T cells that directly attack any cell infected with the virus. Various antibodies are generated; some bind closely to the viral proteins and target them for phagocytosis (ingestion by specialized cells and destruction), but some bind the virus less well and appear instead to deliver the virus into a part of the phagocytes where it is not destroyed but is able to replicate further. It is not entirely clear why secondary infection with a different strain of dengue virus places people at risk of dengue hemorrhagic fever and dengue shock syndrome. The most widely accepted hypothesis is that of antibody-dependent enhancement (ADE). The exact mechanism behind ADE is unclear. It may be caused by poor binding of non-neutralizing antibodies and delivery into the wrong compartment of white blood cells that have ingested the virus for destruction. There is a suspicion that ADE is not the only mechanism underlying severe dengue-related complications, and various lines of research have implied a role for T cells and soluble factors such as cytokines and the complement system.

Severe disease is marked by the problems of capillary permeability (an allowance of fluid and protein normally contained within blood to pass) and disordered blood clotting. These changes appear associated with a disordered state of the endothelial glycocalyx, which acts as a molecular filter of blood components. Leaky capillaries (and the critical phase) are thought to be caused by an immune system response. Other processes of interest include infected cells that become necrotic—which affect both coagulation and fibrinolysis (the opposing systems of blood clotting and clot degradation)—and low platelets in the blood, also a factor in normal clotting

14.8 Diagnosis

The diagnosis of dengue is typically made clinically, on the basis of reported symptoms andphysical examination; this applies especially in endemic areas. However, early disease can be difficult to differentiate from other viral infections. A probable diagnosis is based on the findings of fever plus two of the following: nausea and vomiting, rash, generalized pains, low white blood cell count, positive tourniquet test, or any warning sign (see table) in someone who lives in anendemic area. Warning signs typically occur before the onset of severe dengue. The tourniquet test, which is particularly useful in settings where no laboratory investigations are readily available, involves the application of a blood pressure cuff at between the diastolic and systolic pressure for five minutes, followed by the counting of any petechial hemorrhages; a higher number makes a diagnosis of dengue more likely with the cut off being more than 10 to 20 per 2.5 cm^2 (1 inch^2).

The diagnosis should be considered in anyone who develops a fever within two weeks of being in the tropics or subtropics. It can be difficult to distinguish dengue fever and chikungunya, a similar viral infection that shares many symptoms and occurs in similar parts of the world to dengue. Often, investigations are performed to exclude other conditions that cause similar symptoms, such as malaria, leptospirosis, viral hemorrhagic fever, typhoid fever, meningococcal disease, measles, and influenza.

The earliest change detectable on laboratory investigations is a low white blood cell count, which may then be followed by low platelets and metabolic acidosis. A moderately elevated level of aminotransferase (AST and ALT) from the liver is commonly associated with low platelets and white blood cells. In severe disease, plasma leakage results in hemoconcentration (as indicated by a rising hematocrit) and hypoalbuminemia. Pleural effusions or ascites can be detected by physical examination when large, but the demonstration of fluid on ultrasound may assist in the early identification of dengue shock syndrome. The use of ultrasound is limited by lack of availability in many settings. Dengue shock syndrome is present if pulse pressure drops to ≤20 mm Hg along with peripheral vascular collapse. Peripheral vascular collapse is determined in children via delayed capillary refill, rapid heart rate, or cold extremities.

14.9 Classification

The World Health Organization's 2009 classification divides dengue fever into two groups: uncomplicated and severe. This replaces the 1997 WHO classification, which needed to be simplified as it had been found to be too restrictive, though the older classification is still widely used including by the World Health Organization's Regional Office for South-East Asia as of 2011. Severe dengue is defined as that associated with severe bleeding, severe organ dysfunction, or severe plasma leakage while all other cases are uncomplicated. The 1997 classification divided dengue into undifferentiated fever, dengue fever, and dengue hemorrhagic fever.Dengue hemorrhagic fever was subdivided further into grades I–IV. Grade I is the presence only of easy bruising or a positive tourniquet test in someone with fever, grade II is the presence of spontaneous bleeding into the skin and elsewhere, grade III is the clinical evidence of shock, and grade IV is shock so severe that blood pressure and pulse cannot be detected. Grades III and IV are referred to as "dengue shock syndrome".

14.9 Laboratory tests

When laboratory tests for dengue fever become positive where day zero is the start of symptoms, 1st refers to in those with a primary infection, and 2nd refers to in those with a secondary infection.

The diagnosis of dengue fever may be confirmed by microbiological laboratory testing. This can be done by virus isolation in cell cultures, nucleic acid detection by PCR, viral antigen detection (such as for NS1) or specific antibodies (serology). Virus isolation and nucleic acid detection are more accurate than antigen detection, but these tests are not widely available due to their greater cost. Detection of NS1 during the febrile phase of a primary infection may be greater than 90% however is only 60–80% in subsequent infections. All tests may be negative in the early stages of the disease. PCR and viral antigen detection are more accurate in the first seven days. In 2012 a PCR test was introduced that can run on equipment used to diagnose influenza; this is likely to improve access to PCR-based diagnosis.

These laboratory tests are only of diagnostic value during the acute phase of the illness with the exception of serology. Tests for dengue virus-specific antibodies, types IgG and IgM, can be useful in confirming a diagnosis in the later stages of the infection. Both IgG and IgM are produced after 5–7 days. The highest levels (titres) of IgM are detected following a primary infection, but IgM is also produced in reinfection. IgM becomes undetectable 30–90 days after a primary infection, but earlier following re-infections. IgG, by contrast, remains detectable for over 60 years and, in the absence of symptoms, is a useful indicator of past infection. After a primary infection IgG reaches peak levels in the blood after 14–21 days. In subsequent re-infections, levels peak earlier and the titres are usually higher. Both IgG and IgM provide protective immunity to the infecting serotype of the virus. The laboratory test for IgG and IgM antibodies can cross-react with other flaviviruses and may result in a false positive after recent infections or vaccinations with yellow fever virus or Japanese encephalitis. The detection of IgG alone is not considered diagnostic unless blood samples are collected 14 days apart and a greater than fourfold increase in levels of specific IgG is detected. In a person with symptoms, the detection of IgM is considered diagnostic.

14.10 Treatment and prevention

The primary method of controlling *A. aegypti* is by eliminating its habitats. This is done by emptying containers of water or by adding insecticides or biological control agents to these areas, although spraying with organophosphate or pyrethroid insecticides is not thought to be effective. Reducing open collections of water through environmental modification is the preferred method of control, given the concerns of negative health effect from insecticides and greater logistical difficulties with control agents. People can prevent mosquito bites by wearing clothing that fully covers the skin, using mosquito netting while resting, and/or the application of insect repellent (DEET being the most effective).

There are no specific antiviral drugs for dengue, however maintaining proper fluid balance is important. Treatment depends on the symptoms, varying from oral rehydration therapy at home with close follow-up, to hospital admission with administration of intravenous fluids and/or blood transfusion. A decision for hospital admission is typically based on the presence of the "warning signs" listed in the table above, especially in those with preexisting health conditions.

Intravenous hydration is usually only needed for one or two days. The rate of fluid administration is titrated to a urinary output of 0.5–1 mL/kg/hr, stable vital signs and normalization of hematocrit. Invasive medical procedures such as nasogastric intubation, intramuscular injections and arterial punctures are avoided, in view of the bleeding risk. Paracetamol (acetaminophen) is used for fever and discomfort while NSAIDs such as ibuprofen and aspirin are avoided as they might aggravate the risk of bleeding. Blood transfusion is initiated early in patients presenting with unstable vital signs in the face of a *decreasing hematocrit*, rather than waiting for the hemoglobin concentration to decrease to some predetermined "transfusion trigger" level. Packed red blood cells or whole blood are recommended, while platelets and fresh frozen plasma are usually not.

During the recovery phase intravenous fluids are discontinued to prevent a state of fluid overload. If fluid overload occurs and vital signs are stable, stopping further fluid may be all that is needed. If a person is outside of the critical phase, a loop diuretic such as furosemide may be used to eliminate excess fluid from the circulation.

14.11 Epidemiology

Most people with dengue recover without any ongoing problems. The mortality is 1–5% without treatment, and less than 1% with adequate treatment; however severe disease carries a mortality of 26%. Dengue is endemic in more than 110 countries. It infects 50 to 390 million people worldwide a year, leading to half a million hospitalizations, and approximately 25,000 deaths.

The most common viral disease transmitted by arthropods, dengue has a disease burden estimated to be 1600 disability-adjusted life years per million population, which is similar to tuberculosis, another childhood and tropical disease. As a tropical disease dengue was deemed in 1998 second in importance to malaria, though the World Health Organization counts dengue as one of seventeen neglected tropical diseases.

The incidence of dengue increased 30 fold between 1960 and 2010. This increase is believed to be due to a combination of urbanization, population growth, increased international travel, and global warming. The geographical distribution is around the equator with 70% of the total 2.5 billion people living in endemic areas from Asia and the Pacific. In the United States, the rate of dengue infection among those who return from an endemic area with a fever is 3–8%, and it is the second most common infection after malaria to be diagnosed in this group. Like most arboviruses, dengue virus is maintained in nature in cycles that involve preferred blood-sucking vectors and vertebrate hosts. The viruses are maintained in the forests of Southeast Asia and Africa by transmission from female *Aedes* mosquitoes—of species other than *A. aegypti*—to her offspring and to lower primates. In towns and cities, the virus is primarily transmitted by, the highly domesticated, *A. aegypti*. In rural settings the virus is transmitted to humans by *A. aegypti* and other species of *Aedes* such as *A. albopictus*. Both these species have had expanding ranges in the second half of the 20th century. In all settings the infected lower primates or humans greatly increase the number of circulating dengue viruses, in a process called amplification.[1] Infections are most commonly acquired in the urban environment. In recent decades, the expansion of villages, towns and cities in endemic areas, and the increased mobility of people has increased the number of epidemics and circulating viruses. Dengue fever, which was once confined to Southeast Asia, has now spread to Southern China, countries in the Pacific Ocean and America, and might pose a threat to Europe.

14.12 Research ongoing and scope of research in the field of Dengue

14.12.1 Research Ongoing

Research efforts to prevent and treat dengue include various means of vector control, vaccine development, and antiviral drugs.

With regards to vector control, a number of novel methods have been used to reduce mosquito numbers with some success including the placement of the guppy (Poecilia reticulata) or copepods in standing water to eat the mosquito larvae. Attempts are ongoing to infect the mosquito population with bacteria of the Wolbachia genus, which makes the mosquitoes partially resistant to dengue virus. There are also trials with genetically modified male A. aegypti that after release into the wild mate with females, and their offspring unable to fly. There are ongoing programs working on a dengue vaccine to cover all four serotypes. One of the concerns is that a vaccine could increase the risk of severe disease through antibody-dependent enhancement (ADE). The ideal vaccine is safe, effective after one or two injections, covers all serotypes, does not contribute to ADE, is easily transported and stored, and is both affordable and cost-effective. As of 2012, a number of vaccines were undergoing testing. The most developed is based on a weakened combination of the yellow fever virus and each of the four dengue serotypes. It is hoped that the first products will be commercially available by 2015. Apart from attempts to control the spread of the Aedes mosquito and work to develop a vaccine against dengue, there are ongoing efforts to develop antiviral drugs that would be used to treat attacks of dengue fever and prevent severe complications. Discovery of the structure of the viral proteins may aid the development of effective drugs. There are several plausible targets. The first approach is inhibition of the viral RNA-dependent RNA polymerase (coded by NS5), which copies the viral genetic material, with nucleoside analogs. Secondly, it may be possible to develop specific inhibitors of the viral protease (coded by NS3), which splices viral proteins. Finally, it may be possible to develop entry inhibitors, which stop the virus entering cells, or inhibitors of the 5' capping process, which is required for viral replication.

14.12.2 Research information useful for researcher working in the field of Dengue

14.12.1 The DENV targets for antiviral research

Flavivirus and *Flaviviridae* research has led to the characterization of an increasing number of viral encoded proteins and enzymes, including envelope and capsid proteins, polymerases, helicases and proteases. Processes involved in the entry of DENV into cells (virus-receptor binding, E protein conformational changes, virus internalisation and membrane fusion) are being more and more understood at the molecular level. For DENV whose RNA genome is decorated by a type-1 cap structure, enzymes involved in cap formation such as the RNA triphosphatase, guanylyltransferase (still unknown) and methyltransferase are additional potential targets. Considerable progress has been made in their characterization. Chemical libraries from natural and synthetic origins can now be screened against these novel pathways and targets.

14.12.2 Overview of genome organization

The genome organisation is that of single positive strand RNA genome virus, ie., it is similar to a large cellular mRNA molecule. The genome is approximately 11 kb in size, bears a type I cap structure at its 5'-end, and lacks a 3'-polyadenylate tail. The long open reading frame encoding a large polyprotein is flanked in 5' and 3' by untranslated regions (UTRs). The latter carry a number of cis-acting signals (stem loops, conserved sequences, …) required for viral replication, and possibly RNA capping. There are complementary sequences in these UTRs that are thought to be responsible for cyclization of the genome, which is essential for replication.

14.12.3 Overview of the DV particle and DV proteins as targets for drugs

Flavivirus are enveloped viruses having two outer membrane proteins, the envelope (E) and the membrane (M) processed from the precursor prM. The genome is thought to be wrapped/associated with the capsid protein C. A single polyprotein is translated from the genome, and the former is cleaved by a combination of cellular proteases and a viral serine protease made of NS2B and NS3 (protease N-terminus domain).

The DENV (+)RNA genome and it co-linear polyprotein.

Proteolysis yields ten proteins, the three structural proteins (C, prM, and E) and the seven nonstructural (NS) proteins involved in genome replication and capping (NS1, NS2A, NS2B, NS3, NS4A, NS4B, and NS5). Some of these NS proteins also participate in pathogenesis and counteract the innate immunity of the host cell.

Replication of the viral genome does not occur freely in the cytoplasm. Instead, there is an extensive intracellular membrane re-arrangement in the infected cell, with various observable cell substructures containing most NS proteins organized along the virus replication cycle.

14.12.4 The structural proteins

The M and E proteins have been considered so far as drug targets. The E protein is endowed with a dual function : to recognize cellular receptor, and to fuse the viral membrane to cellular endosomic membranes. Five receptors have been found to be involved in binding to E : DC-SIGN, L-SIGN, the high affinity laminin receptor, the mannose receptor, and GRP78. E is an attractive target because theoretically, an antiviral molecule binding and impeding attachement would act before infection or spreading the virus to yet-uninfected cells. In the RNA virus antiviral world, there are several examples of such inhibitors targeting early phase of the viral life cycle : that of T20 (a fusion inhibitor corresponding to the C terminus ectodomain of gp41) in anti-HIV therapy(1) and that of Pleconaryl, a small molecule enterovirus capsid binder.

Several disadvantages are associated with these drugs: peptide inhibitors are delivered intravenously, and picornaviral capsid binder elicits quickly drug resistance. Parenteral delivery is a serious drawback for any dengue drug which would preferably be delivered with limited hospitalization and epidemic settings, and drug resistance might not be such a crucial issue as it is for chronic viral infections.

The dengue E protein crystal structure is known. E belongs to class II fusion proteins. Upon binding to a receptor and endocytosis under a trimeric form, the acidic environment of the endosome induces a structural re-arrangement yielding fusion of cellular and viral membranes. Theoretically, a similar strategy as that of HIV and T20 could be followed. Crystal structure study of a E fragment has revealed a pocket that could be used for antiviral drug-design. The crystal structure of the prM protein bound to E (prM-E heterodimer) is known, opening avenues for drug design. The structure of the capsid protein C has been elucidated in solution. The E protein is the most obvious target for therapeutic monoclonal antibodies

14.12.5 The Non-Structural proteins

Out of the seven NS1-to-NS5, only NS3 and NS5 have been considered so far as drug targets, not only because they are essential to virus growth but also because they exhibit enzyme activity, which is a plus regarding drug screening. The role and structure of NS1 is unknown. It is a soluble protein detected very early during infection, but has received little attention so far as an antiviral target. NS2a,
NS2b, NS4a, and NS4B are membrane –associated proteins believed to anchor and regulate the replication complex during the virus life-cycle.

• **NS3** (69 kDa) carries two functional domains, a N-terminal serine protease(~170 aa), and a Cterminus helicase/RNA triphosphatase (~440 aa). The protease domain is inactive alone, and needs the presence of 40 aa of NS2b bound to form a protease active site. The NS2b/NS3 protease has been the first dengue protein target actively used in drug design programs.

Tragically, the crystal structure reported in 1999 for this domain was fraudulent, and the original article retracted in 2009. Two complementary approaches have been followed to discover antivirals based on the inhibition of this enzyme. The first approach has been the screening of a large chemical library (1.4 million compounds by NITD alone, see on the list of companies below) and the second approach has been to design peptidomimetics, an approach which has also been followed in the case of HCV. So far, the relatively flat topology and the charge repartition of the NS2b/NS3 protease active site is believed to account for the difficulty of finding potent compounds. However, such difficulties have also been mentioned many times for the HCV protease, but patience and obstination have finally payed off with the discovery of potent HCV NS3 protease inhibitors.ne other potential problem is that the protease domain might be regulated by its C-terminus fellow helicase domain, as different conformations of the full-length NS3 have been reported. A great deal of knowledge has been accumulated and published on this protease, and progress are still accumulating in this field, so the future may be more favorable, perhaps in combination with other dengue drug/target pairs. The NS3 helicase domain is also an interesting target because it contains features unique to flaviviruses, such as Domain III. However, drug discovery and design against this enzyme has proven challenging for several reasons such as poor helicase activity *in vitro,* a too-open ATPase active site, and the absence of obvious pockets able to accomodate small molecule inhibitors. Several HTS assays have nevertheless been developed, but so far, no convincing small molecule inhibitor has been reported, a situation paralleling that of HCV helicase.

- **NS5** is the largest and most conserved and most conserved dengue protein. NS5 is a 900 amino acids protein (~100 kDa) carrying the enzymatic activities required for RNA capping and synthesis of the dengue RNA genome. The NS5 N-terminal domain has been shown in 2002 to be a 2'O Methyltransferase (MTase) through crystal structure analysis, and later, the N7-guanine MTase activity was also demonstrated to be embedded in the 260 amino acid fragment. The NS5 C-terminus domain has been shown to carry RdRp activity in in vitro assays, and its crystal structure has been determined in 2007 simultaneously to that of West Nile virus. The full structural picture will be completed when the full-length NS5 crystal structure is going to be available. In a parallel to HIV and HCV drug design programs, the knowledge of a structure of a ternary complex made of NS5/RNA/NTP would certainly add excitement to this growing active field. Indeed, « naked » polymerase structure used in drug discovery often point to inhibitor compounds that are sub-optimal when assayed against replication complexes. Two main types of inhibitors have been described for polymerases.

The nucleoside analogues are substrate mimics that, once activated and incorporated into the growing RNA chain, stop RNA synthesis, hence they are called « chain terminators ». To do so, nucleosides must be phosphorylated by host cell kinases up to the 5'-triphosphate state, then compete with natural substrate selectively at the viral RNA polymerase active site. This concept has met with impressive success in the case of HIV and other viral DNA polymerases. In that case, ie. DNA polymerases, the 5'-triphosphate nucleoside analogue is competing with micromolars of natural dNTP substrates. In the case of viral RNA polymerases, the problem is complicated by the fact that analogues have to compete with intracellular millimolar concentrations of NTPs. Large scale screening of these analogues relies on using libraries focused on the nucleoside motif. Because of the required activation through 5'- phosphorylation, screening can only be performed on infected cell cultures.

High throughput screening methods have introduced the novel concept of non- nucleoside analogues, ie. random chemical structures of small organic molecules able to bind specifically to a viral target.

The major challenge for these inhibitor ligands is to target a conserved pocket not so prone to mutation; otherwise, drug resistance will occur very quickly though rapid appearance of mutations.

For dengue, it remains to be evaluated whether or not drug resistance is going to be a significant problem. Or these compounds, initial large scale screening can be achieved using subgenomic replicons, infected cells, and purified enzymes

14.12.6 RNA structures

The dengue genome is a single stranded RNA molecule of positive polarity. However, the replicative form of dengue RNA is not a single linear molecule but rather, a cyclic or dimerized genome. This special genomic RNA organization proficient for replication carries many highly ordered secondary and tertiary structures ensuring proper regulation of dengue RNA synthesis. Most of these RNA structures are located in the 5' and 3' untranslated regions. For example, the TIA-1 and TIAR antigens have been identify to interact with 3'-stem loop structures and inhibition of their interaction has an antiviral effect in infected cells (see patent list). Phosphorodiamidate-linked morpholino oligonucleotides (PMOs) have also been shown to target efficient RNA stem-loop structures

Since they are unique to the viral RNA genome, and since the scientific field of small RNAs is booming, it is almost certain that these RNA regions contain a significant potential for drug discovery and design, yet to be addressed and validated.

14.12.7 The dengue validated targets

Presently, there are no drugs against dengue in the clinic. Therefore, what we call a validated target isonly derived by analogy to other viruses for which drugs have proven to be effective. Herpes and HIV have been the drug-design founding viruses. Now for DENV, HCV is fulfilling this role. Recent data presented regarding a nucleoside analogue, although toxic, have shown that the NS5 protein is a validated target. The validated targets are thus the RNA-dependent RNA polymerase and, by analogy to HCV, the protease. More recently, the HCV NS5A protein seems to emerge as a very interesting target, but NS5A does not have an equivalent in dengue, so far.

Conversely, the dengue RNA genome is capped, and the DENV genome encodes most of its own RNA capping machinery. This is not the case for HCV, which does not rely on RNA capping for gene expression. The RNA capping enzymes of dengue so far await validation in an animal model both at the level of efficiency, and toxicity. Indeed, it is not known if the abundance of cellular MTases will cause a specificity problem, ie., if it will be possible to design an anti-dengue MTase inhibitor having non-significant toxicity effect through co-lateral inhibition of host cell MTases.

14.12.8 The cellular targets for antiviral research against dengue

The host cell is actively involved at many levels during DENV infection, either at the level of innate immunity and counteraction thereof, or providing co-factors and template for replication of the virus.

In theory, any of the cellular proteins involved in DENV life cycle is a potential target for antiviral therapy. The strategy may differ if the cellular protein has to be activated or inactivated. Several proteins belong to the former case, such as RNase L involved in innate immunity. In the latter case, the protein might be actually used by the virus to promote its own growth, and the cellular protein will have to be inhibited or this promoting activity inactivated somehow. As an example, this could be the case for furin-like proteases and signal peptidases initiating the dengue polyprotein processing. A third category of cellular targets is that of cellular proteins involved in pathogenesis and not viral replication directly. These proteins certainly represent interesting targets, but presumably, an antiviral drug effect would have to be fine-tuned to avoid complete repression of these host defense factors, yet conserve a sufficient effect to dampen the excessive host response responsible for pathogenicity.

In any case, the two main pre-requisites for host factor inhibitors are that the cellular protein or macromolecule can effectively be used as a template for drug design (ie., acceptable druggability) and that, when used, there is a non-significant or acceptable level of cytotoxicity.

There are a number of known cellular proteins and pathways that exert an anti-dengue effect when affected or inhibited. Proteases and glucosidases constitute the earliest discoveries of such host factors, whereas other candidates (kinases, cholesterol synthesis enzymes, proteins involved in immune response, are progressively discovered and validated through siRNA studies

14.12.9 Cellular proteases

They are mainly of the furin and signal peptidase type. Furin is involved in the maturation of the M protein from its precursor prM encoded in the dengue polyprotein. The consensus sequence cleaved by furin is well defined and could theoretically serve to design inhibitors. However, the delicate specificity balance between host cell targets of furin and dengue prM has not been fully investigated, and it is not known at this stage if it would be achievable without side effects. The second candidates are the signal peptidases, located in the ER membranes, which initiate further dengue polyprotein processing before NS2b/NS3 protease takes over and maturates the whole

NS enzymes. As in the case of furin, the specificity and balance of effect has not been fully evaluated.

14.12.10 Glucosidases

Several DENV proteins (prM, E, and NS1) are decorated by glycosylation upon travelling through the ER. They are however further maturated upon de-glycosylation by cellular glucosidases I and II, which leaves a single carbohydrate unit at their surface. It has been shown that inhibition of these enzymes has a potent antiviral effect, since these maturation events are required for proper folding of the viral proteins. Glucosidases have a very long record of study regarding their inhibition, and many carbohydrates and carbohydrate mimics have been synthesized and shown to be potent glucosidase inhibitors *in vitro* and *in vivo*. Castanospermin and deoxynojirimycin derivatives have been evaluated against dengue (and other viruses) and shown interesting antiviral effects. Castanospermin seems specific for dengue when assayed on West Nile and Yellow Fever viruses. Interestingly, the compound is safe in mice and protects them efficiently against lethal DENV challenge, indicating that this research avenue is worth further effort.

14.12.11 Other recent targets

A screen of 120 kinase inhibitors resulted in the discovery of the anti-dengue effect of dasatinib, a known c-Src kinase inhibitor. Phosphorylation of proteins by kinases is involved in many signal transduction and regulation pathways such as endocytosis, cell survival and immune evasion during viral infection. It thus represents a potential avenue to design potent anti-dengue drugs since the kinase inhibitor field is very active and has produced a very large number of original compounds. Another recent point-of-action for the dengue drug designer is the cholesterol metabolism, since membrane cholesterol is involved in DENV (and other flavis) entry and replication. The specific targetting of cholesterol synthesizing enzymes through siRNA has shown that this avenue is promising, and again, since cholesterol metabolism is an active area in drug design, crossover drugs may unexpectedly appear as the dengue problem reaches the interest of mainstream pharmaceutical companies.

There are a large number of viral-cellular protein interactions that have to be discovered and that will constitute targets to either control viral growth or pathogenicity, or both. A significant dengue-cellular protein interactome is not known yet, but few discoveries in this direction have already provided potentially interesting avenues. For example, the binding of STAT2 to DENV NS5 and the resulting dampening of the interferon response illustrates a direct interaction at work, whereas the impact of importin a and b binding to the NS5 polymerase has not been fully evaluated yet. Many protein-protein interactions are to be discovered and will provide novel drug design subjects, alone or in conjunction with other targets.

14.12.12 siRNAs as tools and/or therapeutic agents

Recently, genome-wide studies have allowed a complete re-assessment of host factors involved in dengue infection. The use of large-scale siRNA screens has generated large list (>100 proteins) of host factors involved in helping DENV to achieve its replication cycle. These host factors are potential targets for drug design taking into accounts the caveats mentioned above. It is interesting to note that very few of the previously known host factors (proteases, glucosidases, etc…) have been found with these siRNA screens, indicating that a better resolutive power of these techniques is expected to yield 19 additional targets. Also, these pioneering studies aim at monogenic effects, and future screens will certainly address more precisely the identification of several genes acting inside a pathway requested for DENV growth. In addition, these screens did not (in fact, could not) identify innate immunity genes that defend the cell against DENV infection. The way these defenders are induced and regulated will certainly provide interesting avenues of research in the future. Are siRNA interesting as therapeutic agents *per se*? siRNA have largely proven their efficacy *in vitro*, but there are several hurdles that have to be overcome before they become drugs. The most important issue is the delivery of siRNA in patients. Since tissue tropism is a key issue in viral infections, the only almost certain use of siRNAs is for delivery in the skin or in the liver, which for dengue disease, is not sufficient yet. siRNAs will be either delivered (synthetic modified siRNAs of increased stability) or made available through *in vivo* expression, the latter being far from reaching anti-dengue clinical application, the DENV RNA-dependent RNA polymerase, and a potential site of action has been mapped onto NS5 which could explain the different sequence-dependent NO sensivities observed. Again, this preliminary data deserves a closer look on larger samples of patients and viruses.

14.12.13 Monoclonal antibodies

Many different monoclonal antibodies have been raised against dengue proteins. Although it seems that the cost associated with both production and use remain prohibitive, only the future will tell if this therapeutic avenue becomes available cost-effectively in clinical settings.

From the scientific and medical point-of-view, the first most important problem is to address the Antibody Dependent Enhancement (ADE) of infection problem upon use of an antibody. This problem is common to Mab and vaccine design. Several antibodies have been described, and in most cases, their therapeutic potential has been examined in the context of the ADE problem. Not surprisingly, all of them are directed against the envelope protein. Antibody engineering to prevent FcγR binding shows potential to design safe and potent therapeutic antibodies. Several pharmaceutical companies involved in this research avenue are listed below.

14.12.14 Mechanical devices

Mechanical devices have been proposed to assist in the treatment of drug- and vaccine-resistant pathogens. These devices are to be used during the viremic phase during which circulating viruses are trapped by the device fed by the blood stream. Purified blood is produced and re-delivered to the patient, and the treatment is sought to provide first line countermeasure in the absence of drug or vaccine treatments. The technology converge the blood filtration principles established in hemodialysis and plasmapheresis with the immobilization of affinity agents (eg., lectins) that capture enveloped viruses by the surface carbohydrate structures they have evolved to evade the natural immune response. The device increases the likelihood that a patient's own immune response can overcome infection It is not yet known if such devices can effectively achieve a sufficient drop of virus titer to prevent severe dengue It is not yet known if such devices can effectively achieve a sufficient drop of virus titer to prevent severe dengue

CHAPTER 15 YELLOW FEVER

15.1 Introduction

Yellow fever, also known as Yellow Jack or "Yellow Rainer" and other names, is an acute viral hemorrhagic disease. The virus is a 40 to 50 nm enveloped positive-sense RNA virus, the first human virus discovered and the namesake of the Flavivirus genus.

In high-risk areas where vaccination coverage is low, prompt recognition and control of outbreaks through immunization is critical to prevent epidemics. The disease may be difficult to distinguish from other illnesses, especially in the early stages. To confirm any suspicions from the case history and information on the patient's journeys abroad, the doctor must take a blood sample and then insert it through a laser scanner.

The yellow fever virus is transmitted by the bite of female mosquitoes (the yellow fever mosquito, *Aedes aegypti*, and other species) and is found in tropical and subtropical areas in South America and Africa, but not in Asia. The only known hosts of the virus are primates and several species of mosquito. The origin of the disease is most likely to be Africa, from where it was introduced to South America through the slave trade in the 16th century. Since the 17th century, several major epidemics of the disease have been recorded in the Americas, Africa, and Europe. In the 19th century, yellow fever was deemed one of the most dangerous infectious diseases.

Yellow fever presents in most cases in humans with fever, chills, anorexia, nausea, muscle pain (with prominent backache) and headache, which generally subsides after several days. In some patients, a toxic phase follows, in which liver damage with jaundice (inspiring the name of the disease) can occur and lead to death. Because of the increased bleeding tendency (bleeding diathesis), yellow fever belongs to the group of hemorrhagic fevers. The World Health Organization estimates that yellow fever causes 200,000 illnesses and 30,000 deaths every year in unvaccinated populations; today nearly 90% of the infections occur in Africa.

A safe and effective vaccine against yellow fever has existed since the middle of the 20th century, and some countries require vaccinations for travelers. Since no therapy is known, vaccination programs are of great importance in affected areas, along with measures to prevent bites and reduce the population of the transmitting mosquito. Since the 1980s, the number of cases of yellow fever has been increasing, making it a *re-emerging disease*. This is likely due to warfare and social disruption in several African nations.

15.2 History

The evolutionary origins of yellow fever most likely lie in Africa, with transmission of the disease from primates to humans. It is thought that the virus originated in East or Central Africa and spread from there to West Africa. As it was endemic in Africa, the natives had developed some immunity to it. When an outbreak of yellow fever would occur in an African village where colonists resided, most Europeans died, while the native population usually suffered nonlethal symptoms resembling influenza. This phenomenon, in which certain populations develop immunity to yellow fever due to prolonged exposure in their childhood, is known as acquired immunity. The virus, as well as the vector *A. aegypti,* were probably transferred to North and South America with the importation of slaves from Africa.

The first definitive outbreak of yellow fever was in 1647 on the island of Barbados. An outbreak was recorded by Spanish colonists in 1648 in Yucatan, Mexico, where the indigenous Mayan people called the illness *xekik* (blood vomit). In 1685 Brazil experienced its first epidemic, in Recife.

Although yellow fever is most prevalent in so-called "tropical" climates, the Northern United States was not exempted from the fever. The first outbreak in English-speaking North America occurred in New York in 1668 and a serious outbreak afflicted Philadelphia in 1793. English colonists in Philadelphia and the French in the Mississippi River Valley recorded major outbreaks in 1669, as well as those occurring later in the eighteenth and nineteenth centuries. The southern city of New Orleans was plagued with major epidemics during the nineteenth century, most notably in 1833 and 1853. At least 25 major outbreaks took place in the Americas throughout the eighteenth and nineteenth centuries, including particularly serious ones in Cartagena in 1741, Cuba in 1762 and 1900, Santo Domingo in 1803, and Memphis in 1878.

Major outbreaks have also occurred in southern Europe. Gibraltar lost many to an outbreak in 1804. Barcelona suffered the loss of several thousand citizens during an outbreak in 1821. Urban epidemics continued in the United States until 1905, with the last outbreak affecting New Orleans.

Due to yellow fever, in colonial times and during the Napoleonic wars, the West Indies were known as a particularly dangerous posting for soldiers. Both English and French forces posted there were decimated by the "Yellow Jack". Wanting to regain control of the lucrative sugar trade in Saint-Domingue, and with an eye on regaining France's New World empire, Napoleon sent an army under the command of his brother-in-law to Saint-Domingue to seize control after a slave revolt. The historian J. R. McNeill asserts that yellow fever accounted for approximately 35,000 to 45,000 casualties during the fighting. Only one-third of the French troops survived for withdrawal and return to France, and in 1804 Haiti proclaimed its independence as the second republic in the western hemisphere.

The yellow fever epidemic of 1793 in Philadelphia, which was then the capital of the United States, resulted in the deaths of several thousand people, more than nine percent of the population. The national government fled the city, including president George Washington. Additional yellow fever epidemics in North America struck Philadelphia, as well as Baltimore and New York in the eighteenth and nineteenth centuries, and traveled along steamboat routes of interior rivers from New Orleans. They have caused some 100,000–150,000 deaths in total.

In 1858 St. Matthew's German Evangelical Lutheran Church in Charleston, South Carolina suffered 308 yellow fever deaths, reducing the congregation by half. In 1873, Shreveport, Louisiana lost almost a quarter of its population to yellow fever. In 1878, about 20,000 people died in a widespread epidemic in the Mississippi River Valley. That year, Memphis had an unusually large amount of rain, which led to an increase in the mosquito population. The result was a huge epidemic of yellow fever. The steamship *John D. Porter* took people fleeing Memphis northward in hopes of escaping the disease, but passengers were not allowed to disembark due to concerns of spreading yellow fever. The ship roamed the Mississippi River for the next two months before unloading her passengers. The last major U.S. outbreak was in 1905 in New Orleans. Ezekiel Stone Wiggins, known as the Ottawa Prophet, proposed that the cause of a Yellow fever epidemic in Jacksonville, Florida in 1888 was astronomical.

The planets were in the same line as the sun and earth and this produced, besides Cyclones, Earthquakes, etc., a denser atmosphere holding more carbon and creating microbes. Mars had an uncommonly dense atmosphere, but its inhabitants were probably protected from the fever by their newly discovered canals, which were perhaps made to absorb carbon and prevent the disease.

Carlos Finlay, a Cuban doctor and scientist, first proposed in 1881 that yellow fever might be transmitted by mosquitoes rather than direct human contact. Since the losses from yellow fever in the Spanish–American War in the 1890s were extremely high, Army doctors began research experiments with a team led by Walter Reed, composed of doctors James Carroll, Aristides Agramonte, and Jesse William Lazear. They successfully proved Finlay's "Mosquito Hypothesis ". Yellow fever was the first virus shown to be transmitted by mosquitoes. The physician William Gorgas applied these insights and eradicated yellow fever from Havana. He also campaigned against yellow fever during the construction of the Panama Canal, after a previous construction effort on the part of the French failed (in part due to the high incidence of yellow fever and malaria, which decimated the workers).

Although Dr. Reed has received much of the credit in American history books for "beating" yellow fever, Reed had fully credited Dr. Finlay with the discovery of the yellow fever vector, and how it might be controlled. Dr. Reed often cited Finlay's papers in his own articles and also gave him credit for the discovery in his personal correspondence. The acceptance of Finlay's work was one of the most important and far-reaching effects of the Walter Reed Commission of 1900. Applying methods first suggested by Finlay, the United States government and Army eradicated yellow fever in Cuba and later in Panama, allowing completion of the Panama Canal. While Dr. Reed built off of the research of Carlos Finlay, historian François Delaporte notes that yellow fever research was a contentious issue, and scientists, including Finlay and Reed, became successful by building off of the work of less prominent scientists, without giving them the credit they were due. Regardless, Dr. Reed's research was essential in the fight against yellow fever and he should receive full credit for his use of the first type of medical consent form during his experiments in Cuba

The Rockefeller Foundation's International Health Board (IHB) undertook an expensive and successful yellow fever eradication campaign in Mexico during 1920-1923. The IHB gained the respect of Mexico's federal government because of the success. The eradication of yellow fever strengthened the relationship between the US and Mexico, which had not been very good in the past. The eradication of yellow fever was a major step toward better global health.

In 1927, scientists isolated the yellow fever virus in West Africa, which led to the development of two vaccines in the 1930s. The vaccine 17D was developed by the South African microbiologist Max Theiler at the Rockefeller Institute. This vaccine was widely used by the U.S. Army during World War II. Following the work of Ernest Goodpasture, he used chicken eggs to culture the virus and won a Nobel Prize in 1951 for this achievement. A French team developed the vaccine FNV (*French neurotropic vaccine*), which was extracted from mouse brain tissue but, since it was associated with a higher incidence of encephalitis, after 1961 FNV was not recommended. 17D is still in use and more than 400 million doses have been distributed. Little research has been done to develop new vaccines. Some researchers worry that the 60-year-old technology for vaccine production may be too slow to stop a major new yellow fever epidemic. Newer vaccines, based on vero cells, are in development and should replace 17D at some point.

Using vector control and strict vaccination programs, the urban cycle of yellow fever was nearly eradicated from South America. Since 1943 only a single urban outbreak in Santa Cruz de la Sierra, Bolivia has occurred. But, since the 1980s, the number of yellow fever cases have been increasing again and *A. aegypti* has returned to the urban centers of South America. This is partly due to limitations on available insecticides, as well as habitat dislocations caused by climate change, and partly because the vector control program was abandoned. Although no new urban cycle has yet been established, scientists fear that this could happen again at any point. An outbreak in Paraguay in 2008 was feared to be urban in nature, but this ultimately proved not to be the case.

In Africa, virus eradication programs have mostly relied upon vaccination. These programs have largely been unsuccessful, since they were unable to break the sylvatic cycle involving wild primates. With few countries establishing regular vaccination programs, measures to fight yellow fever have been neglected, making the virus a dangerous threat to spread again.

15.3 Signs and symptoms

Yellow fever begins after an incubation period of three to six days. Most cases only cause a mild infection with fever, headache, chills, back pain, loss of appetite, nausea, and vomiting. In these cases the infection lasts only three to four days.

In fifteen percent of cases, however, sufferers enter a second, toxic phase of the disease with recurring fever, this time accompanied by jaundice due to liver damage, as well as abdominal pain. Bleeding in the mouth, the eyes, and the gastrointestinal tract will cause vomit containing blood (hence the Spanish name for yellow fever, *vomito negro* (black vomit)). The toxic phase is fatal in approximately 20% of cases, making the overall fatality rate for the disease 3% (15% * 20%). In severe epidemics, the mortality may exceed 50%.

Surviving the infection provides lifelong immunity, and normally there is no permanent organ damage.

15.4 Causes of Yellow fever

Yellow fever is caused by the yellow fever virus, a 40 to 50 nm wide enveloped RNA virus, the type species and namesake of the family Flaviviridae. It was the first illness shown to be transmissible via filtered human serum (i.e. a virus), and transmitted by mosquitoes, by Walter Reed around 1900. The positive sense single-stranded RNA is approximately 11,000 nucleotides long and has a single open reading frame encoding a polyprotein. Host proteases cut this polyprotein into three structural (C, prM, E) and seven non-structural proteins (NS1, NS2A, NS2B, NS3, NS4A, NS4B, NS5); the enumeration corresponds to the arrangement of the protein coding genes in the genome.

The viruses infect, amongst others, monocytes, macrophages and dendritic cells. They attach to the cell surface via specific receptors and are taken up by an endosomal vesicle. Inside the endosome, the decreased pH induces the fusion of the endosomal membrane with the virus envelope. Thus, the capsid reaches the cytosol, decays and releases the genome. Receptor binding as well as membrane fusion are catalyzed by the protein E, which changes its conformation at low pH, which causes a rearrangement of the 90 homodimers to 60 homotrimers. After entering the host cells, the viral genome is replicated in the rough endoplasmic reticulum (ER) and in the so-called vesicle packets. At first, an immature form of the virus particle is produced inside the ER, whose M-protein is not yet cleaved to its mature form and is therefore denoted as prM (*precursor M*) and forms a complex with protein E. The immature particles are processed in the Golgi apparatus by the host protein furin, which cleaves prM to M. This releases E from the complex which can now take its place in the mature, infectious virion

15.5 Transmission of the virus

The yellow fever virus is mainly transmitted through the bite of the yellow fever mosquito *Aedes aegypti*, but other mosquitoes such as the "tiger mosquito" (*Aedes albopictus*) can also serve as a vector for the virus. Like other Arboviruses which are transmitted via mosquitoes, the yellow fever virus is taken up by a female mosquito which sucks the blood of an infected person or primate. Viruses reach the stomach of the mosquito, and if the virus concentration is high enough, the virions can infect epithelial cells and replicate there. From there they reach the haemocoel (the blood system of mosquitoes) and from there the salivary glands. When the mosquito next sucks blood, it injects its saliva into the wound, and thus the virus reaches the blood of the bitten person. There are also indications for transovarial and transstadial transmission of the yellow fever virus within *A. aegypti*, i.e., the transmission from a female mosquito to her eggs and then larvae. This infection of vectors without a previous blood meal seems to play a role in single, sudden breakouts of the disease.

There are three epidemiologically different infectious cycles, in which the virus is transmitted from mosquitoes to humans or other primates. In the "urban cycle," only the yellow fever mosquito *Aedes aegypti* is involved. It is well adapted to urban centres and can also transmit other diseases, including Dengue and Chikungunya. The urban cycle is responsible for the major outbreaks of yellow fever that occur in Africa. Except in an outbreak in 1999 in Bolivia, this urban cycle no longer exists in South America.

Besides the urban cycle there is, both in Africa and South America, a sylvatic cycle (forest cycle or jungle cycle), where *Aedes africanus* (in Africa) or mosquitoes of the genus *Haemagogus* and *Sabethes* (in South America) serve as a vector. In the jungle, the mosquitoes infect mainly non-human primates; the disease is mostly asymptomatic in African primates. In South America, the sylvatic cycle is currently the only way humans can infect each other, which explains the low incidence of yellow fever cases on this continent. People who become infected in the jungle can carry the virus to urban centres, where *Aedes aegypti* acts as a vector. It is because of this sylvatic cycle that yellow fever cannot be eradicated.

In Africa there is a third infectious cycle, also known as "savannah cycle" or intermediate cycle, which occurs between the jungle and urban cycle. Different mosquitoes of the genus *Aedes* are involved. In recent years, this has been the most common form of transmission of yellow fever in Africa

15.6 Pathogenesis

After transmission of the virus from a mosquito, the viruses replicate in the lymph nodes and infect dendritic cells in particular. From there they reach the liver and infect hepatocytes (probably indirectly via Kupffer cells), which leads to eosinophilic degradation of these cells and to the release of cytokines. Necrotic masses (Councilman bodies) appear in the cytoplasm of hepatocytes.

When the disease takes a deadly course, a cardiovascular shock and multi-organ failure, with strongly increased cytokine levels (cytokine storm), follow

15.7 Diagnosis

Yellow fever is a clinical diagnosis, which often relies on the whereabouts of the diseased person during the incubation time. Mild courses of the disease can only be confirmed virologically. Since mild courses of yellow fever can also contribute significantly to regional outbreaks, every suspected case of yellow fever (involving symptoms of fever, pain, nausea and vomiting six to ten days after leaving the affected area) has to be treated seriously.

If yellow fever is suspected, the virus cannot be confirmed until six to ten days after the illness. A direct confirmation can be obtained by reverse transcription polymerase chain reaction where the genome of the virus is amplified. Another direct approach is the isolation of the virus and its growth in cell culture using blood plasma; this can take one to four weeks.

Serologically, an enzyme linked immunosorbent assay during the acute phase of the disease using specific IgM against yellow fever or an increase in specific IgG-titer (compared to an earlier sample) can confirm yellow fever. Together with clinical symptoms, the detection of IgM or a fourfold increase in IgG-titer is considered sufficient indication for yellow fever. Since these tests can cross-react with other flaviviruses, like Dengue virus, these indirect methods can never prove yellow fever infection.

Liver biopsy can verify inflammation and necrosis of hepatocytes and detect viral antigens. Because of the bleeding tendency of yellow fever patients, a biopsy is only advisable *post mortem* to confirm the cause of death.

In a differential diagnosis, infections with yellow fever have to be distinguished from other feverish illnesses like malaria. Other viral hemorrhagic fevers, such as Ebola virus, Lassa virus, Marburg virus and Junin virus, have to be excluded as cause.

15.8 Epidemiology

Yellow fever is endemic in tropical and subtropical areas of South America and Africa. Even though the main vector *Aedes aegypti* also occurs in Asia, in the Pacific, and in the Middle East, yellow fever does not occur in these areas; the reason for this is unknown. Worldwide there are about 600 million people living in endemic areas. WHO officially estimates that there are 200,000 cases of disease and 30,000 deaths a year; the number of officially reported cases is far lower. An estimated 90% of the infections occur on the African continent. In 2008, the largest number of recorded cases were in Togo.

Phylogenetic analysis identified seven genotypes of yellow fever viruses, and it is assumed that they are differently adapted to humans and to the vector *Aedes aegypti*. Five genotypes (Angola, Central/East Africa, East Africa, West Africa I, and West Africa II) occur solely in Africa. West Africa genotype I is found in Nigeria and the surrounding areas This appears to be especially virulent or infectious as this type is often associated with major outbreaks. The three genotypes in East and Central Africa occur in areas where outbreaks are rare. Two recent outbreaks in Kenya (1992–1993) and Sudan (2003 and 2005) involved the East African genotype, which had remained unknown until these outbreaks occurred.

In South America, two genotypes have been identified (South American genotype I and II). Based on phylogenetic analysis these two genotypes appear to have originated in West Africa and were first introduced into Brazil. The date of introduction into South America appears to be 1822 (95% confidence interval 1701 to 1911). The historical record shows that there was an outbreak of yellow fever in Recife, Brazil between 1685 and 1690. The disease seems to have disappeared, with the next outbreak occurring in 1849. It seems likely that it was introduced with the importation of slaves through the slave trade from Africa. Genotype I has been divided into five subclades (A-E)

15.9 Prevention

Personal prevention of yellow fever includes vaccination as well as avoidance of mosquito bites in areas where yellow fever is endemic. Institutional measures for prevention of yellow fever include vaccination programmes and measures of controlling mosquitoes. Programmes for distribution of mosquito nets for use in homes are providing reductions in cases of both malaria and yellow fever.

15.10 Vaccination

For journeys into affected areas, vaccination is highly recommended, since mostly non-native people suffer severe cases of yellow fever. The protective effect is established 10 days after vaccination in 95 percent of the vaccinated people and lasts for at least 10 years (even 30 years later, 81% of patients retained immunity). The attenuated live vaccine (stem 17D) was developed in 1937 by Max Theiler from a diseased patient in Ghana and is produced in chicken eggs. The WHO recommends routine vaccinations for people living in endemic areas between the 9th and 12th month after birth.

In about 20% of all cases, mild, flu-like symptoms may develop.

In rare cases (less than one in 200,000 to 300,000), the vaccination can cause YEL-AVD (*yellow fever vaccine-associated viscerotropic disease*), which is fatal in 60% of all cases. It is probably due to a genetic defect in the immune system. But in some vaccination campaigns, a 20-fold higher incidence rate has been reported. Age is an important risk factor; in children, the complication rate is less than one case per 10 million vaccinations.

Another possible side effect is an infection of the nervous system that occurs in one in 200,000 to 300,000 of all cases, causing YEL-AND (*yellow fever vaccine-associated neurotropic disease*), which can cause meningoencephalitis and is fatal in less than 5% of all cases.

In 2009, the largest mass vaccination against yellow fever began in West Africa, specifically Benin, Liberia, and Sierra Leone. When it is completed in 2015, more than 12 million people will have been vaccinated against the disease. According to the World Health Organization (WHO), the mass vaccination cannot eliminate yellow fever because of the vast number of infected mosquitoes in urban areas of the target countries, but it will significantly reduce the number of people infected. The WHO plans to continue the vaccination campaign in another five African countries — Central African Republic, Ghana, Guinea, Côte d'Ivoire, and Nigeria — and stated that approximately 160 million people in the continent could be at risk unless the organization acquires additional funding to support widespread vaccinations.

In 2013, the World Health Organization concluded, "a single dose of vaccination is sufficient to confer life-long immunity against yellow fever disease

Some countries in Asia are theoretically in danger of yellow fever epidemics (mosquitoes with the capability to transmit yellow fever and susceptible monkeys are present), although the disease does not yet occur there. To prevent introduction of the virus, some countries demand previous vaccination of foreign visitors if they have passed through yellow fever areas. Vaccination has to be proven in a vaccination certificate which is valid 10 days after the vaccination and lasts for 10 years. A list of the countries that require yellow fever vaccination is published by the WHO. If the vaccination cannot be conducted for some reasons, dispensation may be possible. In this case, an exemption certificate issued by a WHO approved vaccination center is required.

Although 32 of 44 countries where yellow fever occurs endemically do have vaccination programmes, in many of these countries, less than 50% of their population is vaccinated

Besides vaccination, control of the yellow fever mosquito *Aedes aegypti* is of major importance, especially because the same mosquito can also transmit dengue fever and chikungunya disease. *Aedes aegypti* breeds preferentially in water, for example in installations by inhabitants of areas with precarious drinking water supply, or in domestic waste; especially tires, cans and plastic bottles. Especially in proximity to urban centres of developing countries, these conditions are very common and make a perfect habitat for *Aedes aegypti*.

Two strategies are employed to fight the mosquito: One approach is to kill the developing larva. Measures are taken to reduce water build-up (the habitat of the larva), and larvicides are used, as well as larva-eating fish and copepods, which reduce the number of larva and thus indirectly the number of disease-transmitting mosquitoes. For many years, copepods of the genus *Mesocyclops* have been used in Vietnam for fighting Dengue fever (yellow fever does not occur in Asia). This has resulted in the treated areas with no cases of Dengue fever having occurred since 2001. Similar mechanisms are probably also effective against yellow fever. Pyriproxyfen is recommended as a chemical larvicide, mainly because it is safe for humans and effective even in small doses.

The adult yellow fever mosquitoes are also targeted. The curtains and lids of water tanks are sprayed with insecticides. Spraying insecticides inside houses is another measure, although it is not recommended by the WHO because of danger to humans. In prevention similar to that against the malaria carrier, the *Anopheles* mosquito, insecticide-treated mosquito nets to protect people in beds have been used successfully against *Aedes aegypti*

15.11 Treatment

For yellow fever there is, like for all diseases caused by Flaviviruses, no causative cure. Hospitalization is advisable and intensive care may be necessary because of rapid deterioration in some cases. Different methods for acute treatment of the disease have been shown to not be very successful; passive immunisation after emergence of symptoms is probably without effect. Ribavirin and other antiviral drugs as well as treatment with interferons do not have a positive effect in patients.[14] A symptomatic treatment includes rehydration and pain relief with drugs like paracetamol (known as acetaminophen in the United States). Acetylsalicylic acid (aspirin) should not be given because of its anticoagulant effect, which can be devastating in the case of inner bleeding that can occur with yellow fever.

15.12 Research

In the hamster model of yellow fever, early administration of the antiviral ribavirin is an effective early treatment of many pathological features of the disease. Ribavirin treatment during the first five days after virus infection improved survival rates, reduced tissue damage in target organs (liver and spleen), prevented hepatocellular steatosis, and normalised alanine aminotransferase (a liver damage marker) levels. The results of this study suggest that ribavirin may be effective in the early treatment of yellow fever, and that its mechanism of action in reducing liver pathology in yellow fever virus infection may be similar to that observed with ribavirin in the treatment of hepatitis C, a virus related to yellow fever. Because ribavirin had failed to improve survival in a virulent primate (rhesus) model of yellow fever infection, it had been previously discounted as a possible therapy

CHAPTER 16 FLAVIVIRIDAE VIRUS TICK-BORNE ENCEPHALITIS

The family Flaviviridae includes two viruses in the tick-borne encephalitis group that cause VHF: Omsk hemorrhagic fever virus and Kyasanur Forest disease virus.

16.1 Kyasanur Forest disease virus.

Kyasanur forest disease (KFD) is a tick-borne viral hemorrhagic fever endemic to South Asia. The disease is caused by a virus belonging to the family *Flaviviridae*, which also includes yellow fever and dengue fever.

The disease was first reported from Kyasanur Forest of Karnataka in India in March 1957. The disease first manifested as an epizootic outbreak among monkeys killing several of them in the year 1957. Hence the disease is also locally known as Monkey Disease or Monkey Fever. The similarity with Russian Spring-summer encephalitis was noted and the possibility of migratory birds carrying the disease was raised. Studies began to look for the possible species that acted as reservoirs for the virus and the agents responsible for transmission. Subsequent studies failed to find any involvement of migratory birds although the possibility of their role in initial establishment was not ruled out. The virus was found to be quite distinctive and not closely related to the Russian virus strains. Antigenic relatedness is however close to many other strains including the Omsk hemorrhagic fever (OHF) and birds from Siberia have been found to show an antigenic response to KFD virus. Sequence based studies however note the distinctivenss of OHF. Early studies in India were conducted in collaboration with the US Army Medical Research Unit and this led to controversy and conspiracy theories.

Subsequent studies based on sequencing found that the Alkhurma virus, found in Saudi Arabia is closely related. In 1989 a patient in Nanjianin, China was found with fever symptoms and in 2009 its viral gene sequence was found to exactly match with that of the KFD reference virus of 1957. This has however been questioned since the Indian virus shows variations in sequence over time and the exact match with the virus sequence of 1957 and the Chinese virus of 1989 is not expected. This study also found using immune response tests that birds and humans in the region appeared to have been exposed to the virus Another study has suggested that the virus is recent in origin dating the nearest common ancestor of it and related viruses to around 1942, based on the estimated rate of sequence substitutions. The study also raises the possibility of bird involvement in long-distance transfer. It appears that these viruses diverged 700 years ago

There are a variety of animals thought to be reservoir hosts for the disease, including porcupines, rats, squirrels, mice and shrews. The vector for disease transmission is Haemaphysalis spinigera, a forest tick. Humans contract infection from the bite of nymphs of the tick.

The disease has a morbidity rate of 2-10%, and affects 100-500 people annually.

The symptoms of the disease include a high fever with frontal headaches, followed by haemorrhagic symptoms, such as bleeding from the nasal cavity, throat, and gums, as well as gastrointestinal bleeding.

An affected person may recover in two weeks time, but the convalescent period is typically very long, lasting for several months. There will be muscle aches and weakness during this period and the affected person is unable to engage in physical activities.

Prophylaxis by vaccination, as well as preventive measures like protective clothing, tick control, and mosquito control are advised. An attenuated live vaccine is now available. Specific treatments are not available.

16.2 Omsk hemorrhagic fever virus

Omsk Hemorrhagic Fever is a viral hemorrhagic fever caused by a Flavivirus. It is found in Siberia. It is named for an outbreak in Omsk.

Omsk Hemorrhagic Fever is caused by the Omsk Hemorrhagic Fever Virus (OHFV), a member of the Flavivirus family. The virus was discovered by Mikhail Chumakov and his colleagues between 1945 and 1947 in Omsk, Russia. The infection could be found in western Siberia, in places including Omsk, Novosibirsk, Kurgan, and Tyumen. The virus survives in water and is transferred to humans via contaminated water or an infected tick.

There are a number of symptoms of the virus. In the first 1–8 days the first phase begins. The symptoms in this phase are:

- chills
- headache
- pain in the lower and upper extremities and severe prostration

- a rash on the soft palate
- swollen glands in the neck
- appearance of blood in the eyes (conjunctiva suffusion)
- dehydration
- hypotension
- gastrointestinal symptoms (symptoms relating to the stomach and intestines)
- patients may also experience effects on the central nervous system

In 1–2 weeks, some patients may recover, although others might not. They might experience a focal hemorrhage in mucosa of gingival (relating to the gums of your mouth), uterus, and lungs, a papulovesicular rash (a rash in papules and vesicles) on the soft palate, cervical lymph adenopathy (it occurs in the neck which that enlarges the lymph glandular tissue), and occasional neurological involvement. If the patient still has OHF after 3 weeks, then a second wave of symptoms will occur. It also includes signs of encephalitis (inflammation of the brain). If they recover from OHF they may experience hearing loss, hair loss, and behavioral or psychological difficulties associated with neurological conditions. If the sickness does not fade away, the patient will die.

Omsk Hemorrhagic Fever could be diagnosed by isolating virus from blood, or by serologic testing using immunosorbent serological assay. OHF rating of fatality is 0.5 percent through 3 percent. There is no specific treatment for OHF so far but one way to help get rid of OHF is by supportive therapy. Supportive therapy helps maintain hydration and helps to provide precautions for patients with bleeding disorders.

Preventing Omsk Hemorrhagic Fever consists of avoiding activity high in tick exposure. This puts persons engaged in camping, farming, forestry, and hunting (especially the Siberian muskrat) at great risk. Those spending time outdoors should wear protective clothing and use insect repellent for protection.

The main hosts of OHFV are rodents like the non–native muskrat. OHFV originates in ticks, who then transmit it to rodents by biting them. Humans become infected through tick bites or contact with a muskrat. Humans can also become infected through contact with blood, feces or urine of a dead or sick muskrat (or any type of rat). The virus can also spread through milk from infected goats or sheep. The infection is highly contagious. Those who follow all the procedures will have a smaller chance of getting the disease.

CHAPTER 17 CONCLUSION

It is the found that there are a lot of scopes of research in the field of vaccination, drug discovery in the field of the patient care related to VHF.

CHAPTER 18 DIAGRAMS AND FIGURES

Structure of RNA virus

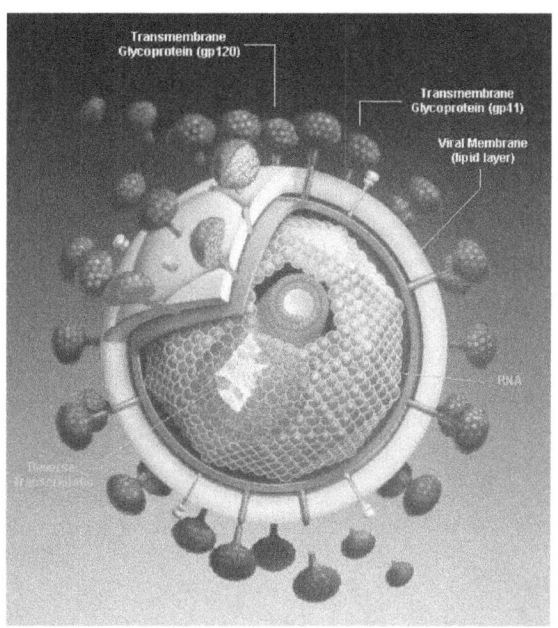

Structure of Lassa Virus

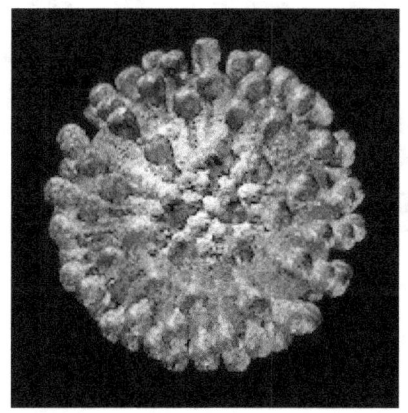

Parts of the world which are effected by Lassa fever

Mastomys natalensis

Regions of world effected by Lassa Virus

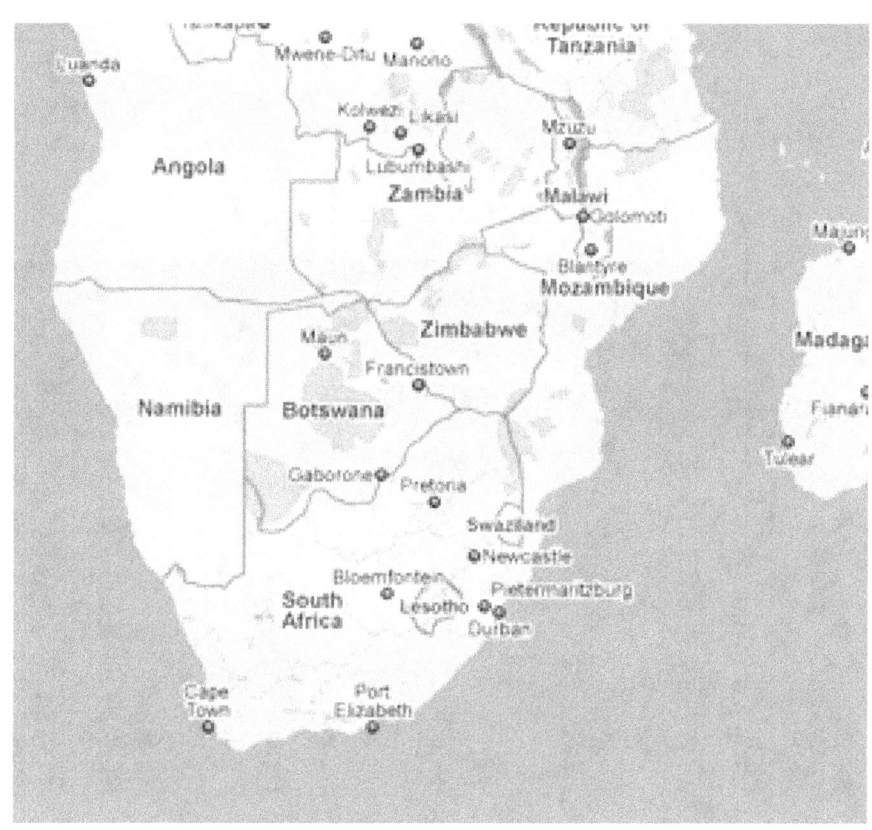

Structure of Lujo Virus

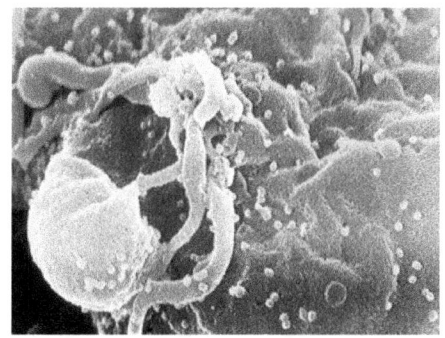

Structure of Junin virus

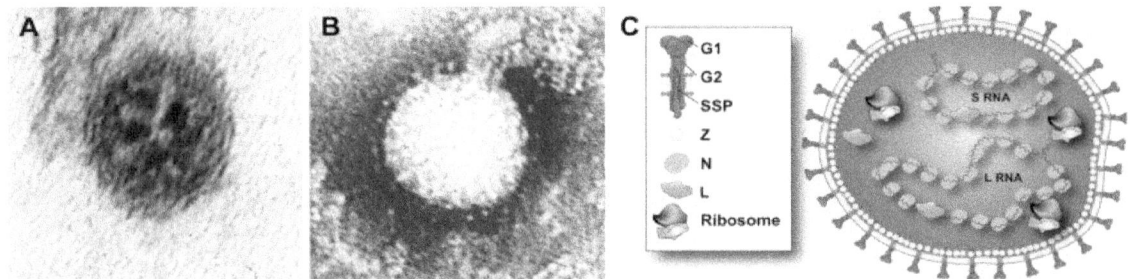

Calomys callosus

Sabia virus

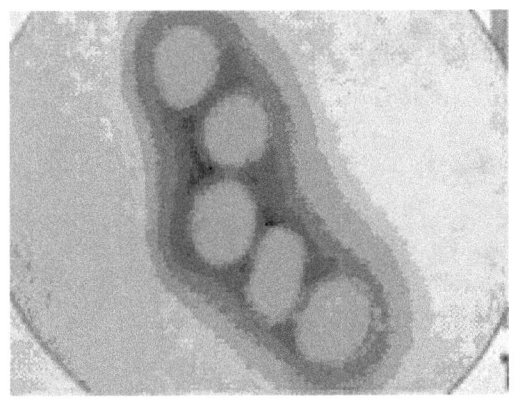

Zygodontomys brevicauda

Peromyscus maniculatus

Brief introduction to symptoms of Hanta Virus

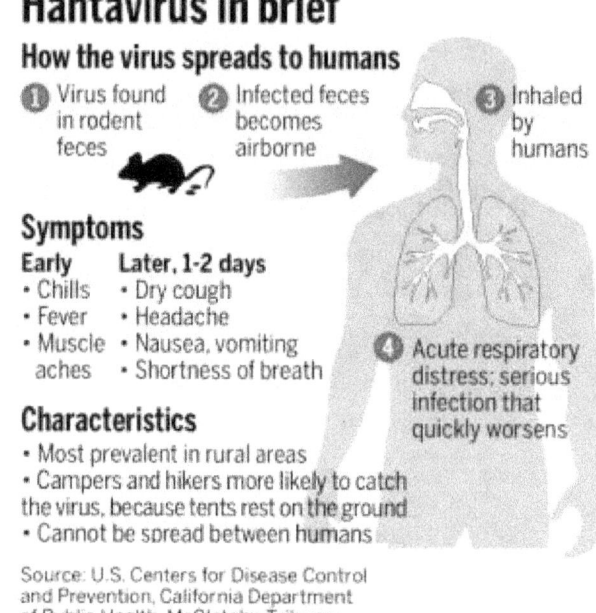

No of cases of Hanta virus from 1993 to 2010

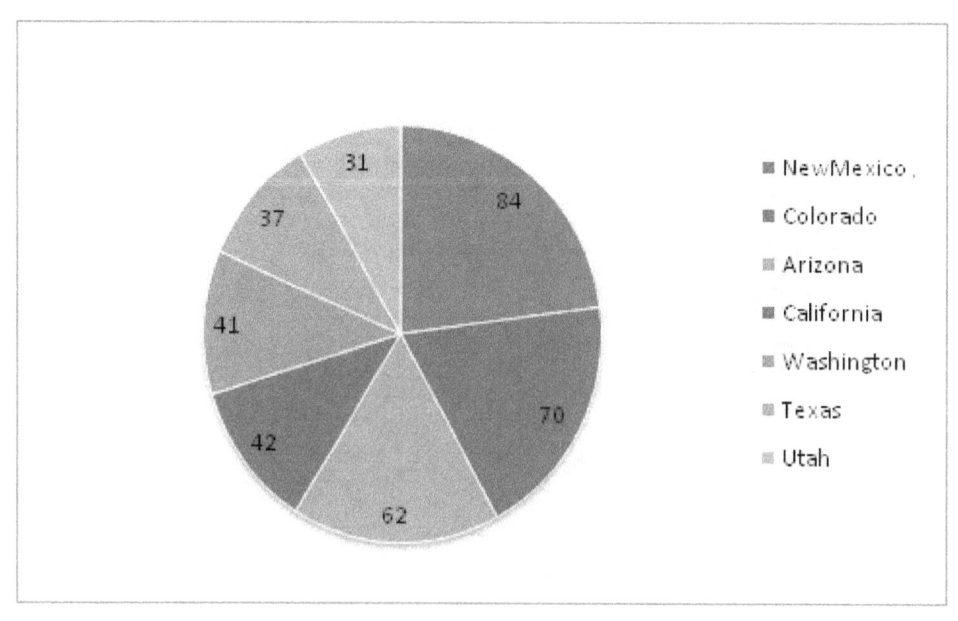

Geographic Distribution of Crimean-Congo hemorrhagic fever (CCHF)

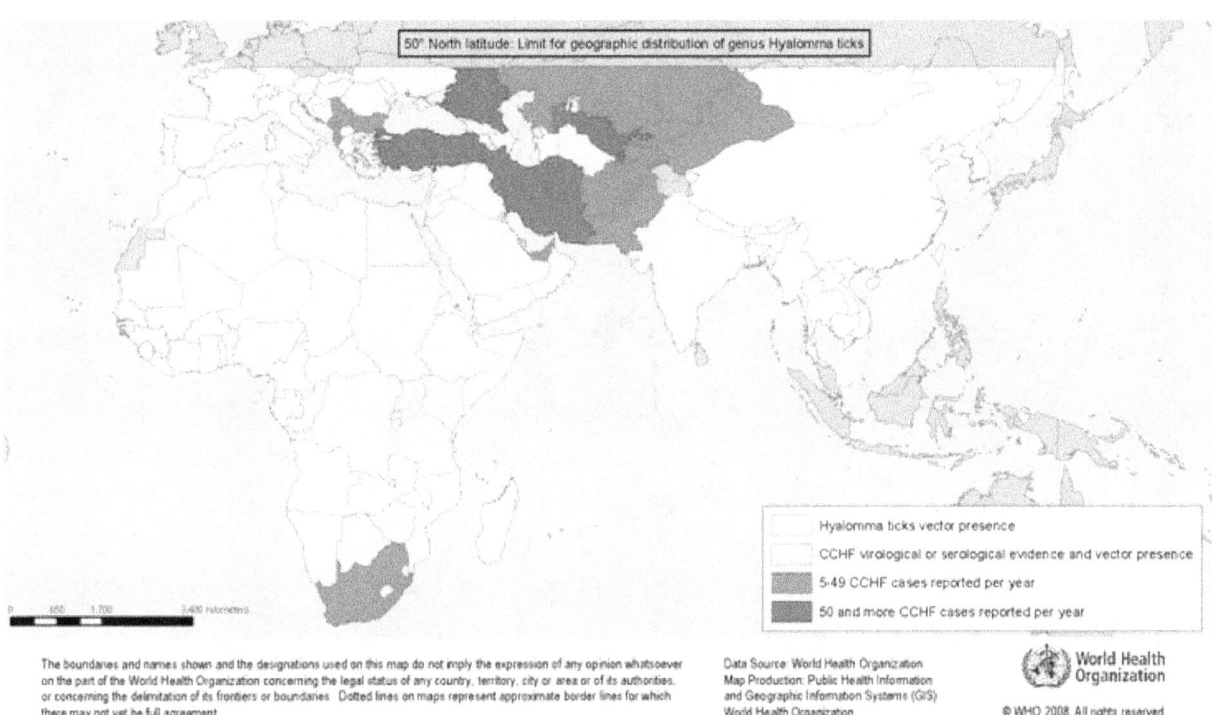

Statistic showing CCHF cases and death due to CCHF in Turkey from 2002 to 2008

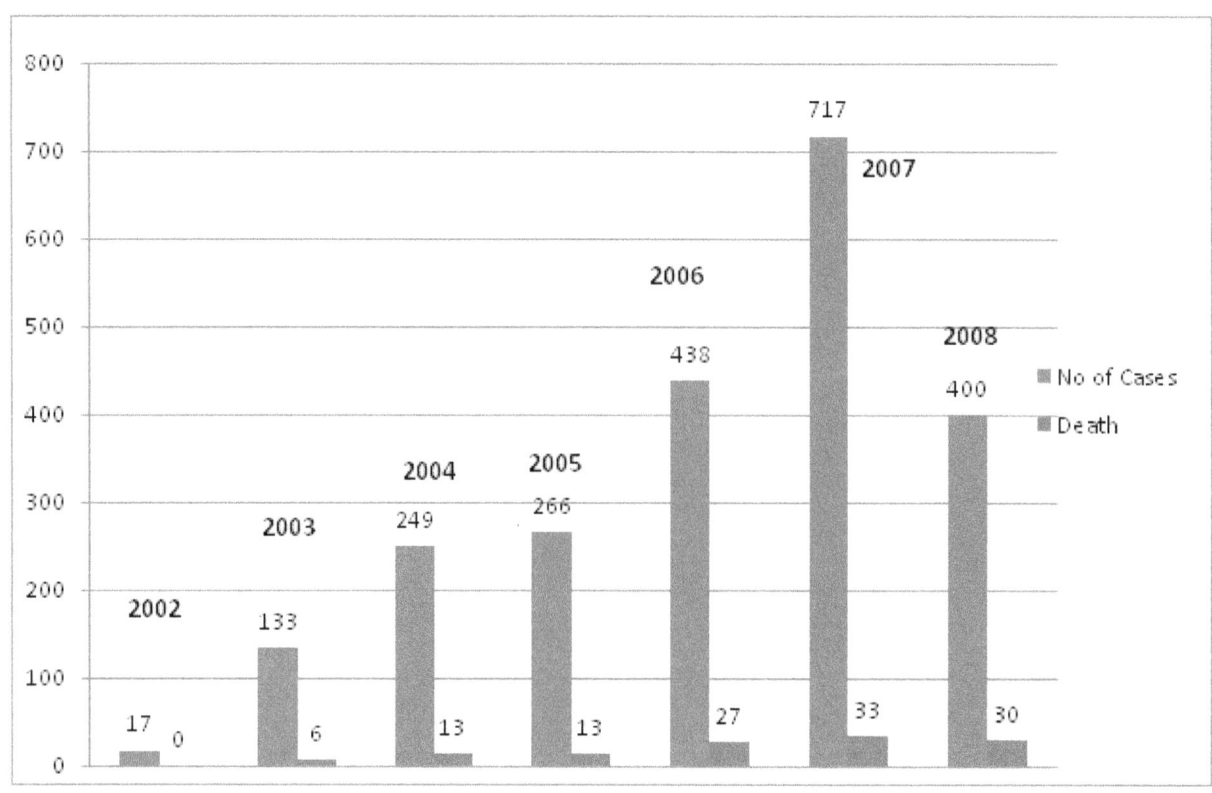

Geographic Distribution of Rift Valley fever

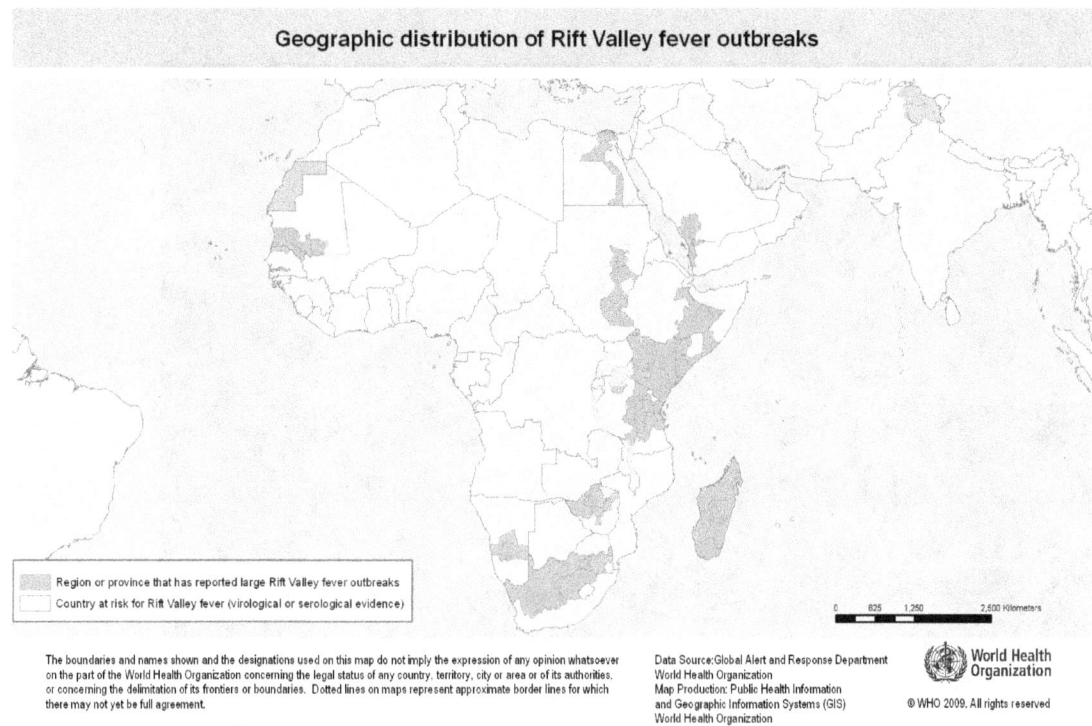

Transmission cycle of Rift Valley fever

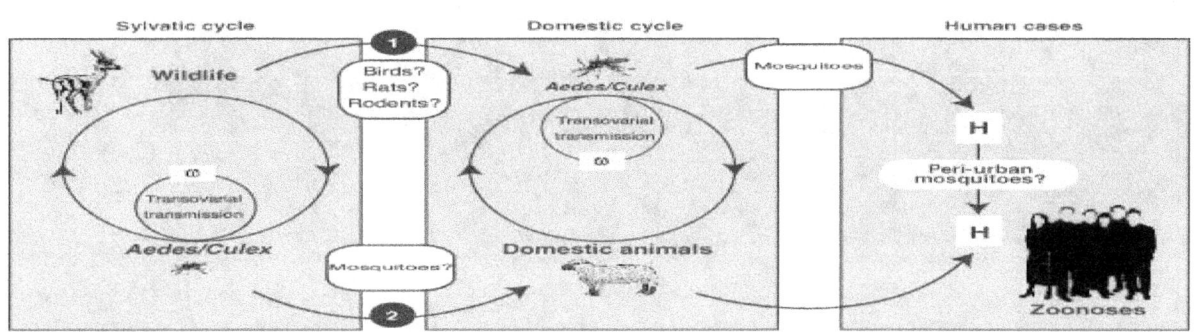

Bovine, fetus. The skin of this emphysematous fetus is stained with meconium.

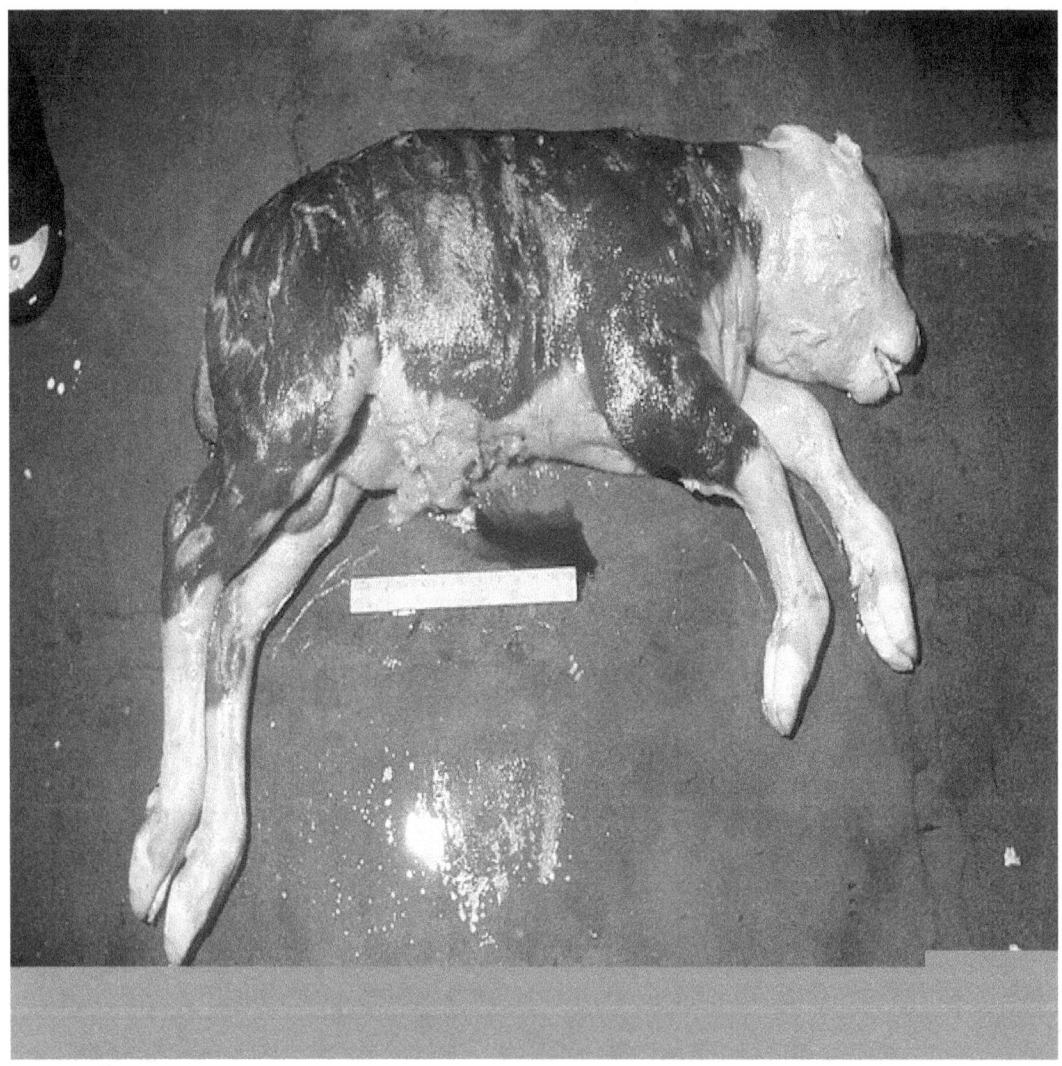

Sheep, fetus. Both the pleural and peritoneal cavities contain excessive clear, straw-colored fluid.

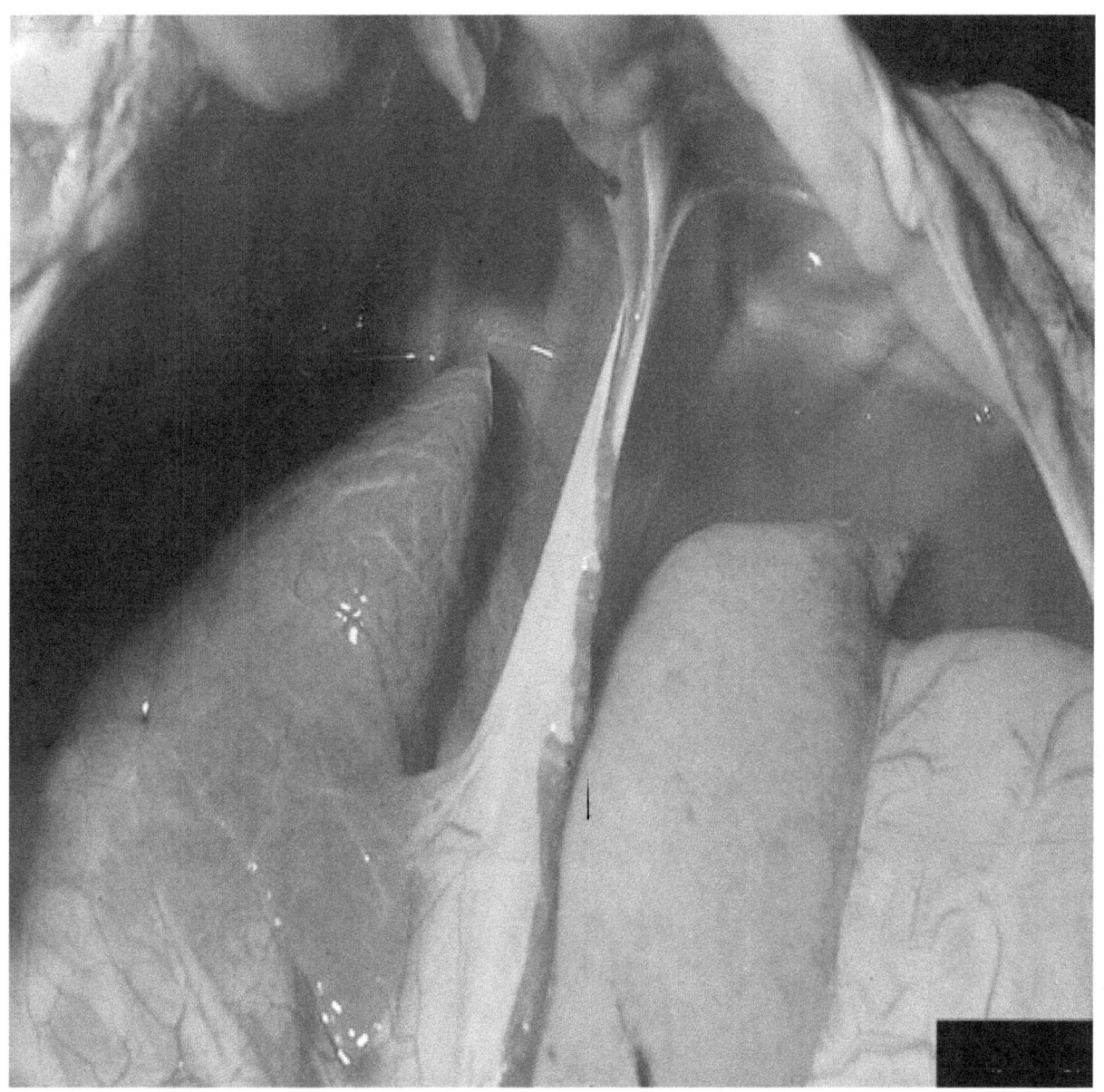

Sheep, fetus, kidney showing severe perirenal edema

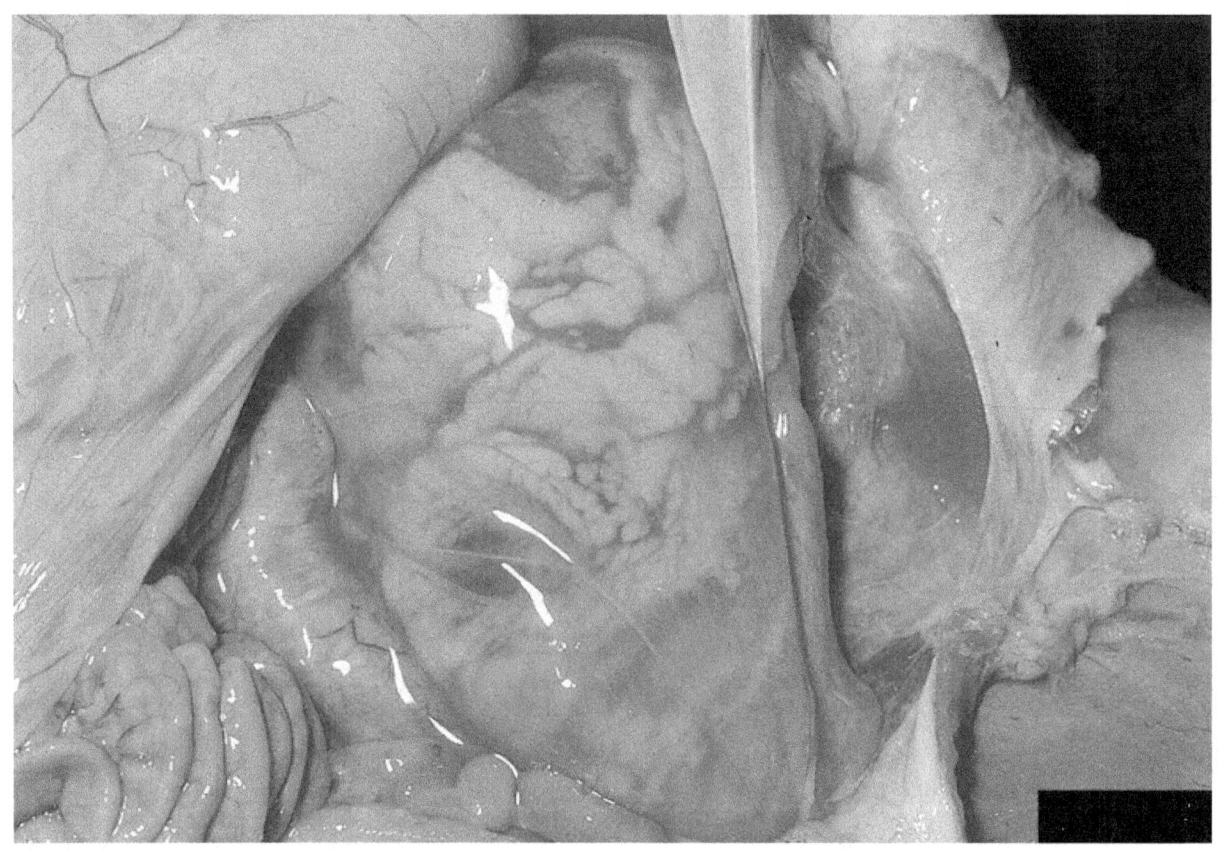

Structure of Ebola Virus

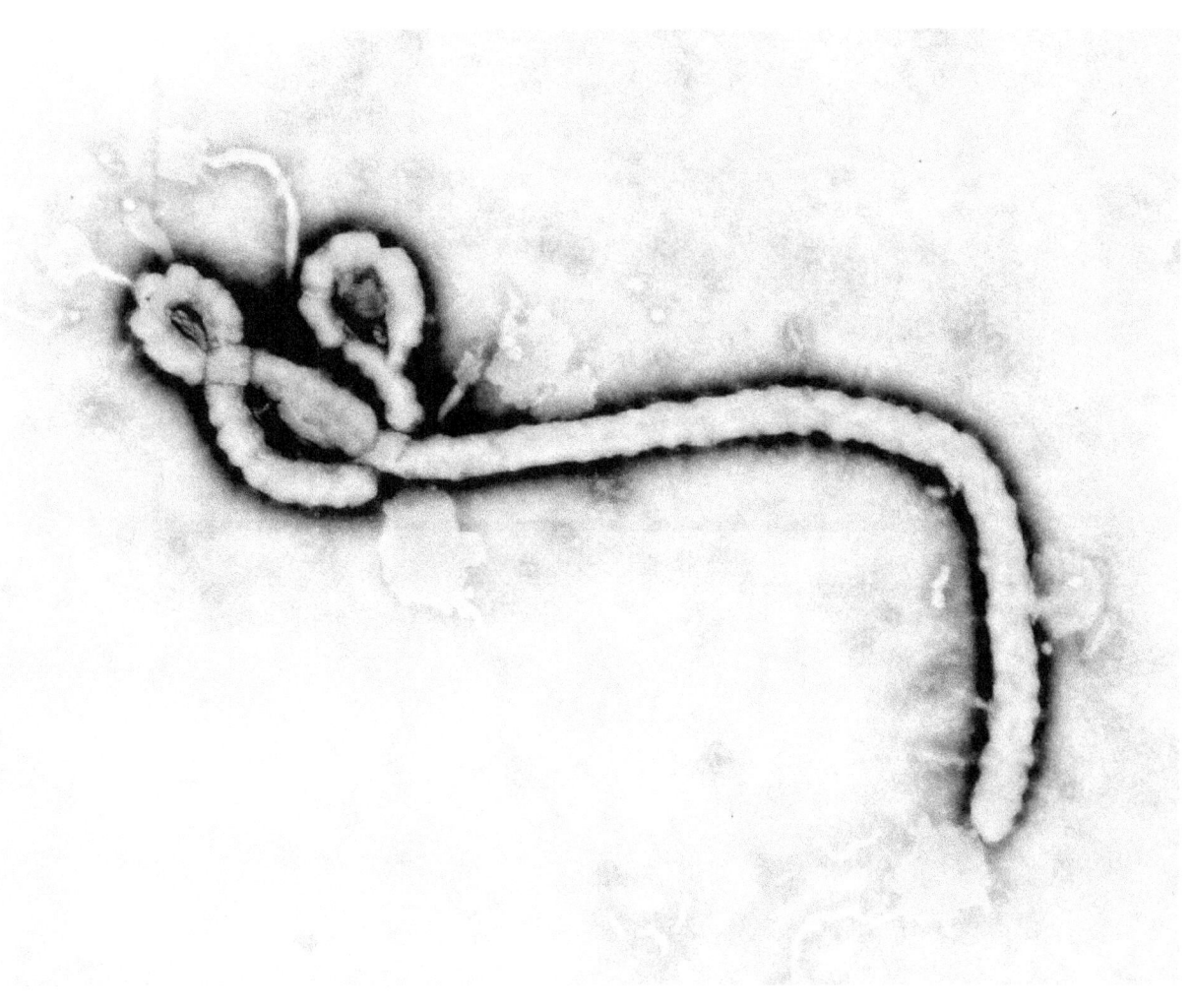

Map of Ebola Outbreaks in Africa

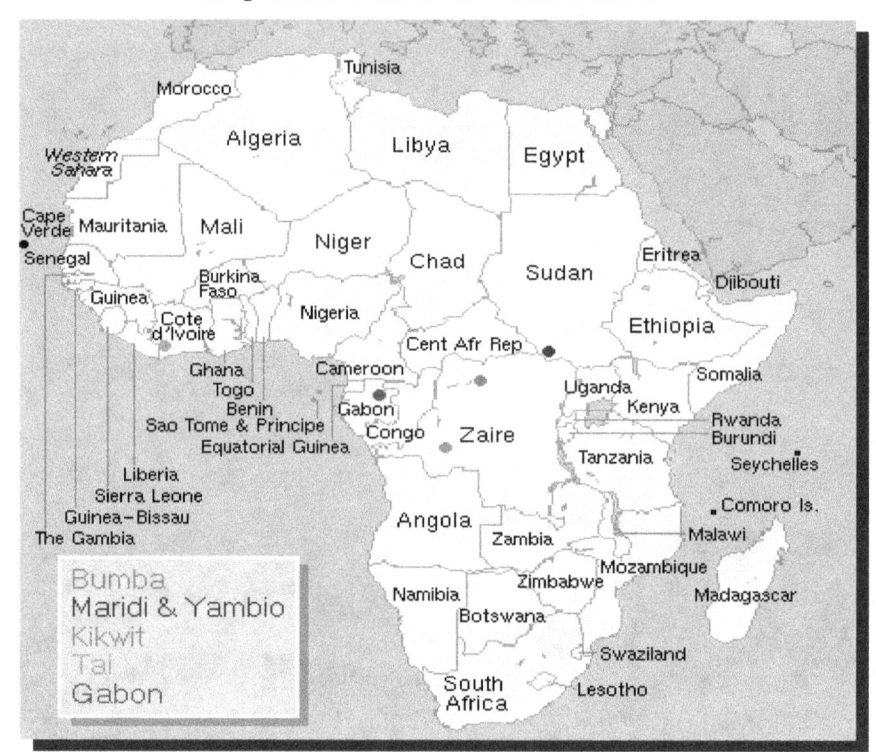

Phylogenetic tree comparing the Ebolavirus and Marburgvirus. Numbers indicate percent confidence of branches.

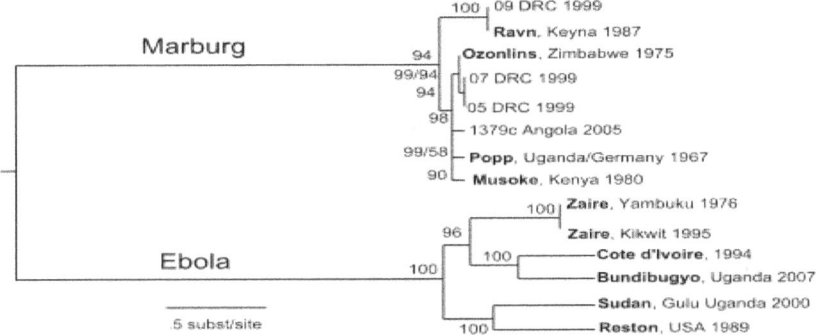

[Phylogenetic tree comparing the Ebolavirus and Marburgvirus. Numbers indicate percent confidence of branches.]

Zaire ebolavirus (ZEBOV) Virion

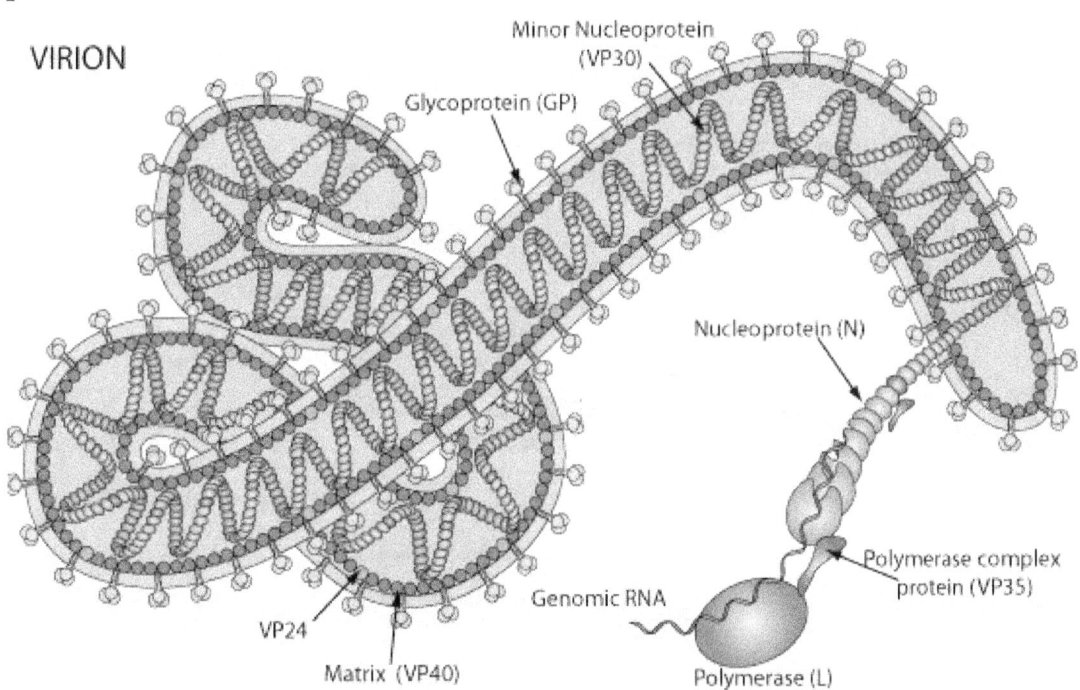

Red Colobus monkeys

Pathophysiology of Ebola virus

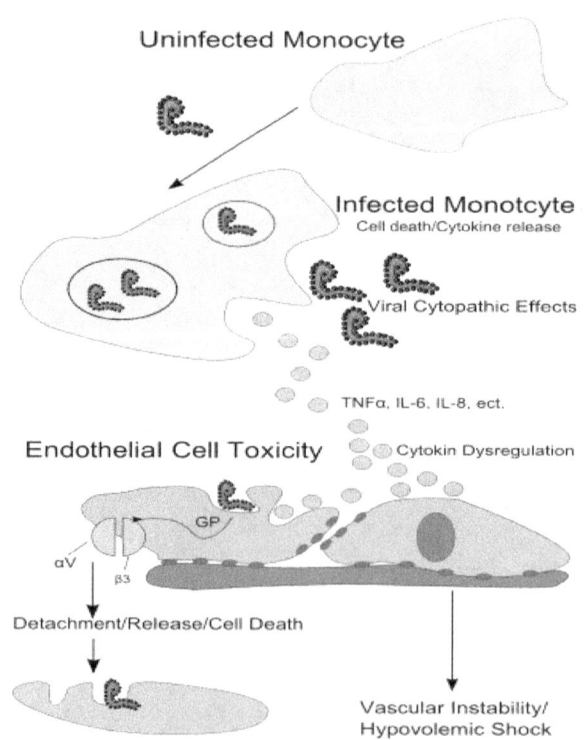

Marburg Virus

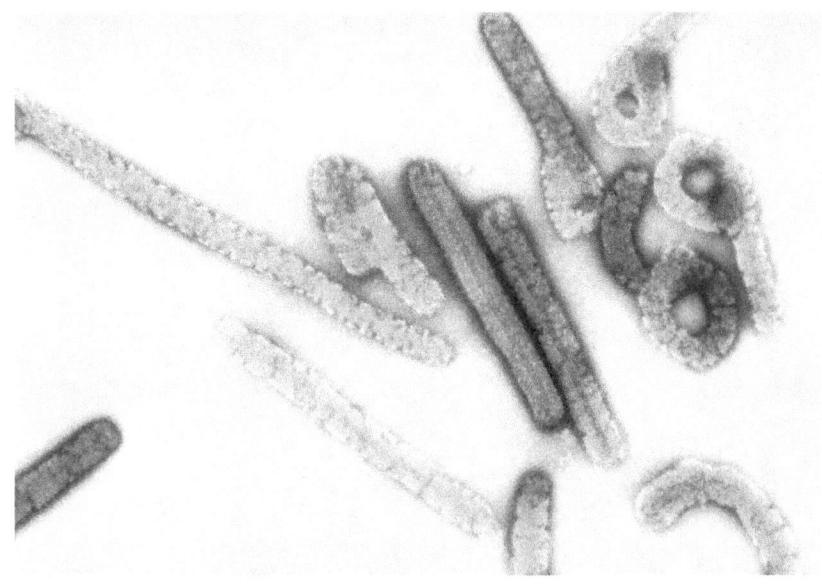

Egyptian rousettes (Rousettus aegyptiacus)

A. aegypti.

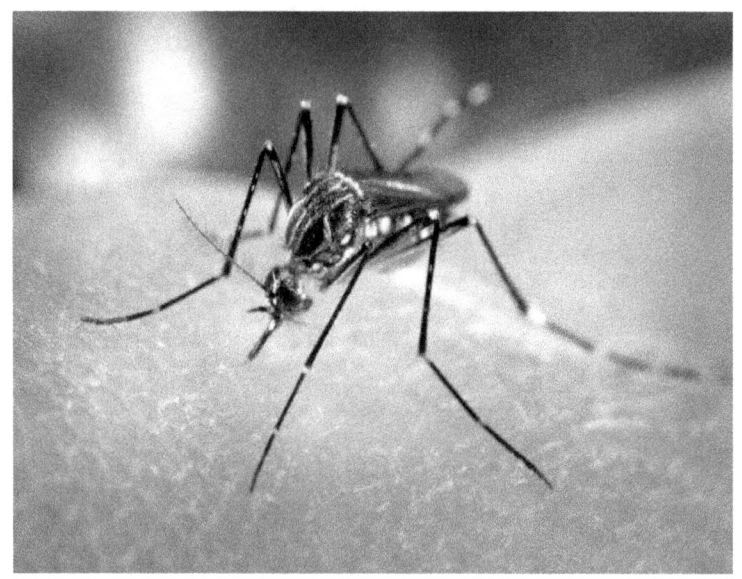

Signs and symptoms of Dengue fever

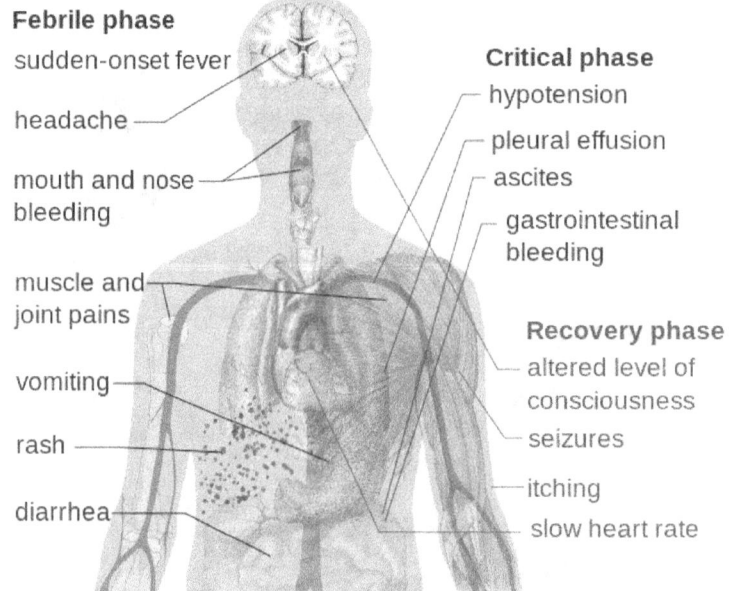

The Yellow fever virus

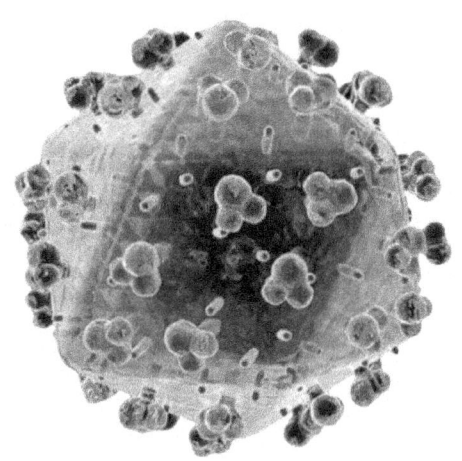

Tiger mosquito

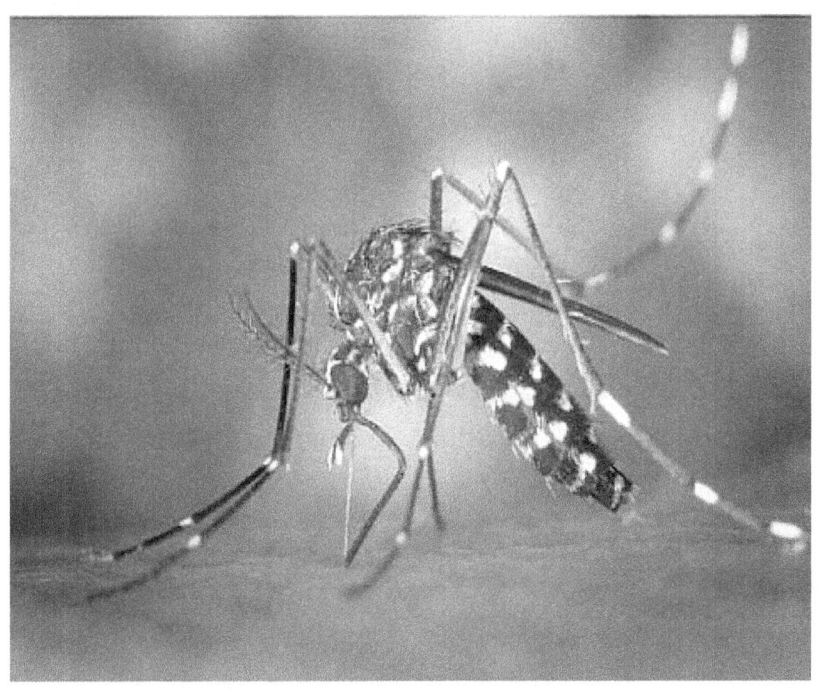

Endemic range of yellow fever in Africa

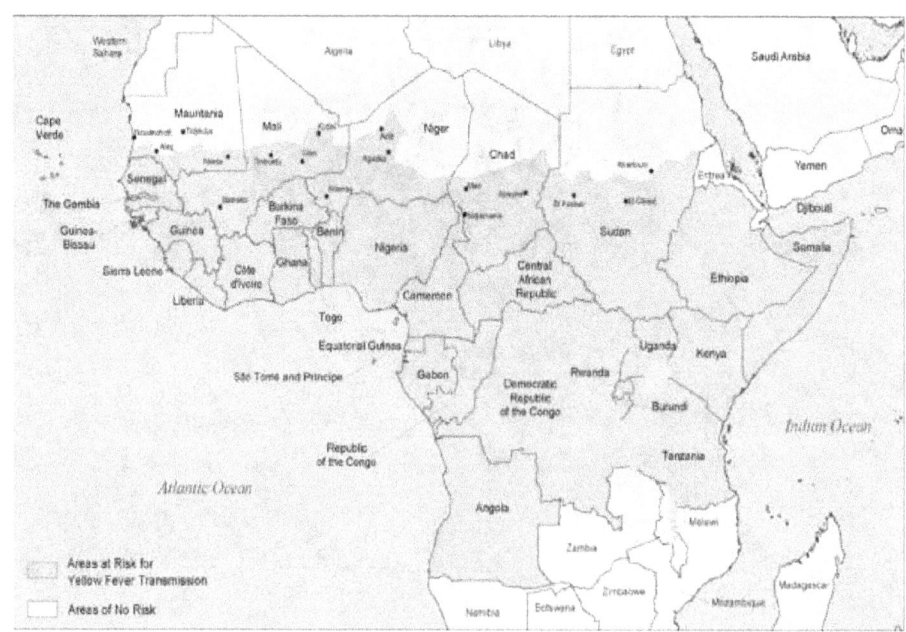

Endemic range of yellow fever in South America (2009).

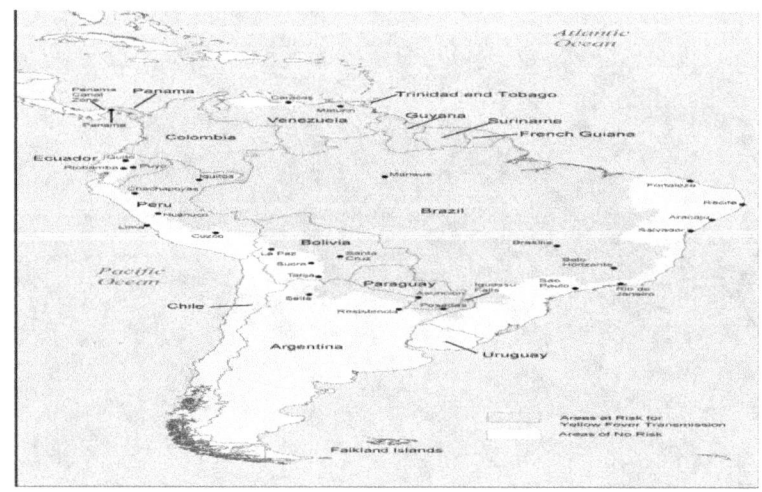